SPACE : UNIVERSE
OF
POLLUTION
Dr. NILESHKUMAR M BARIA

PREFACE

Funda mental ecology is abranch of science aimed at developing mathematical models that could forecast the impact of technogeneous process on the natural environment.the present edition illustrates the methodes and models of fundamental ecology takeing the outer space contamination problem as an example .

Since the first sputnik was launched on 4octomber 1957 and the space era begun mankind was enthusiastic about putting satellite into orbit ,of the wonderful opppurtunities given by the space achievements foe telecommunications,navigations,Earth observations ,weather forecasts.microgravity science and technology,etc. nobody gave thought to a possible negative impact on the space environment,Now it is high time we step aside and look around.

Space activity of mankind generated great deal of orbital debries ,i.e. manmade objects ane their fragments launched into space ,inactive nowadays and not serving any useful purpose. Those objects ,ranging from micron up to decimeters in size,traveling at orbital velocities,remaining in orbot at many years and numbering billions fromed a new media named space debries and become a serious hazard of space flights.collision with metallic particles of debries with 1 centimeter is energeitically equvivalent to a collision with car moving at a speed of 100 km per hour.

Thus this media where in the space satellites operate nowadays should be taken into account ,and its impact on the durability on space missions should be evaluated as it will be affect the reliability of technical systems.That turns out to be of fabulous significance foe upward schemes surrounding constellations of law earth orbiting satellites as a space segment . The prospective 'sky bridge system', a satellite based broadband access the system which will provide high rate internet way in,videoconferesing as well as extra check is supposed to be based on 64 law earth orbit satellites.Developing the ideas of such organizations it is essential to acquire into considerations the space debries environment the system will operate in. (Now the chosen altitude for the sky bridge constellation is 1457 km which practically coinsides with the second greatest of the orbital debries populatios.

Space debries has been studied by leading scientific and research institutions of all the space power since '80s, and its implications has been discussed among the narrow circule of experts of worldwide conferences plus mettings of space agencies from the '90s.but now the emerging constellations of law earth orbiting communication satellites or system providing interactive broadband services and the forthcoming constraction of global space station brought the attention of a wide spectrum of specialists space debries trouble.

The aims of represent eddition are :

- To commence the problem of space debries toa wide circule of people who are consumers of space operating systems;
- To give a decription of the existing level of space contamination plus forecast for prospect development of the orbital debries populations;
- To provide brief coverage of activities on space debries problems that have been in taken and continuing ;
- To highlight the aspect of affecting the growth of space debries poplution in law earth orbits and geosynchronous orbits;
- To present the modern methods of orbital debries evolution modelling ,investigation along with impact hazard assesments ;
- To explain the major hazards for opening space systems and possible preventive measures;
- To converse mitigration strategies ;

The contributors to the book well-known specialists in field of orbital debries learn from Europe,Japan Russia and USA, who have been in this business for many years . most of them experience both working research corporations or space agencies and being professors in universities.they were among the first join the space debries communities,their pioneer publications on the issues are well known.

The book addressed the wide spectrum of readers. An unfamiliar readers find new concept introducing the problem space ecology, new data and examples. A specialists will find actual data on space debries environment.new mathematical models for evolutions ,production and self-production, explanation of the surviveing software ,conception for shield design. A person involving develpoing concept foe multifunctional constellations of law earth orbit satellites will find methods of collison risks assesments depending on altitude and inclinations of orbit,impact risks and evalutations and proposals concerning preventive measures.The book will be suitable as a primary (supplemtary) text for college courses and in-service training programs.

PRICE : 2 Dollar

ACKNOWLEDGEMENT

First of all I pay all my homage and designate my emotions to the lord Shiva whose blessings promote me to accomplish this interesting task. This task requires efforts from many people without them it would not have been completed.

I bow my head in utter humility and complete dedication. It is with great pleasure and with deep sense of gratitude, to my esteemed Father Dr. Madhubhai D. Baria, who has inspired me with keen interest and ever vigilant guidance without which this task could not have been achieved. He has not only guided me but also acted as co-traveler throughout my work. It is the result of his unfailing efforts and meticulous training that today I have reached my destination. The only way to thank his would perhaps to strive to work similarly in years ahead, and continue to chain in succession.

First and foremost, I wish to take this privilege to place on record in a very special and distinctive manner, an expression of deep sense of gratitude and indebtedness to my mother Smt. Dhanuben M. Baria, for her ever inspiring guidance, constant encouragements, friendly support and simplicity has enabled me to complete this work. I am thus highly obliged by her kind support and noble virtues, which will be always cherished by me.

I am extremely grateful to my sister Pareshkumari M. Baria, Associate professor, Government Engineering College, Dahod. For suggesting in work and valuable guidance, stimulates discussion, support and immeasurable help this work till its completion.

EXECUTIVE SUMMERY :

Space pollution refers to the gathering debris in orbit around the Earth, made up of discarded rocket boosters, In the most general sense,

the term space pollution includes both the natural micrometeoroid and man-made orbital debris constituents of the space atmosphere; conversely, as "pollution" is generally considered to indicate a despoiling of the usual environment, space contamination here submits to only man-made orbital debris. Orbital debris poses a hazard to both manned plus unmanned spaceship as well as the earth's inhabitants.

The number, nature, and location of objects superior than 10 cm in size are afforded in the fragmentation debris table and in the image of space debris approximately Earth. Low Earth orbit (LEO) is defined as orbital altitudes below 2,000 km above the earth's surface and is the topic of the representation of space debris just about Earth. Middle Earth orbit (MEO) is the province of the Global Positioning System (GPS) in addition to Russian navigation satellite systems with is located at approximately 20,000-km altitude, whereas the geosynchronous Earth orbit (GEO) "belt" is occupid primarily by communications and Earth—observation payloads about 35,800 km. The mass of objects in these orbital regions are in circular or near-circular orbits about the earth. In dissimilarity, the elliptical orbit class contains rocket bodies left in their transfer (payload delivery) orbits to MEO as well as GEO with scientific, communications, and Earth-observation payloads. Of all objects listed in the

Debris is typically divided into three size ranges, based on the damage it may cause: less than 1 centimeter (cm), 1 to 10 cm, along with superior than 10 cm. Objects less than 1 cm may be shielded against, but they still have the potential to harm the majority satellites. Debris in the 1 to 10 cm range is not shielded against, cannot easily be observed, and could demolish a satellite. lastly, impacts with objects larger than 10 cm can break up a satellite. Of these size ranges, simply objects 10 cm and superior are frequently followed and cataloged by surveillance networks in the United States and the previous Soviet Union. The further populations are estimated statistically through the analysis of returned surfaces (sizes less than 1 mm) or particular measurement campaigns with sensitive radars (sizes larger than 3 mm). Estimates for the populations are just about 30 million debris between 1 mm and 1 cm, over 100,000 debris between 1 and 10 cm, and 8,800 objects larger than 10 cm.

INDEX

Space rubbish, as well standard as orbital waste, space scrap, and space waste, is the group of defunct matter in track about territory. This includes used up sky rocket stages, previous satellites, and waste from collapse, corrosion, and crashs. Since orbits lengthen away from among new spacecraft, garbage may well smash jointly with prepared spacecraft.

As of 2009, about 19,000 pieces of debris superior than 5 cm (2 in) are tracked,[1] with 300,000 pieces superior than 1 cm predictable near exist lower than 2000 km altitude.[1] For contrast, the worldwide Space Station orbits into the 300–400 km array and uniformly the 2009 impact and 2007 antis at test trial happen at linking 800 and 900 km.[1]

Most space debris is smaller than 1 cm (0.4 in), together with dust from solid rocket motors, surface humiliation products such as paint flakes, and frozen coolant droplets free from RORSAT nuclear-powered satellites. Impacts of these particles cause erosive harm, like to sandblasting. Damage can be compact by the accumulation of ballistic shielding to the spacecraft. such as a "Whipple shield", which is use to keep some parts of the International Space Station .still, not all parts of a spacecraft may be protected in this way, e.g. solar panels and optical devices (such as telescopes, or star trackers), and these components are focus to steady wear by debris and micrometeoroids. The change of space debris is greater than meteroids below 2000 km elevation for most sizes circa 2012.[1] Declining risk from space debris larger than 10 cm (4 in) is regularly obtained by maneuvering a spacecraft to avoid a conflict. If a conflict occurs, resulting remains over 1 kg (2 lb)[citation needed] can become an further collision risk.

DEBRIES FLOW

As the occasion of clash is unfair by the number of objects in space, there is a critical density where the conception of new debris is theorized to happen faster than the various usual forces remove them. away from this point, a runaway chain reaction may arise that would quickly increase the number of debris objects in orbit, and thus greatly increase the risk to set satellites. Called the "Kessler syndrome", there is debate if the significant density has previously been reached in certain orbital bands.[2] A escapee Kessler syndrome would render a portion of the helpful polar-orbiting bands complicated to use, and greatly increase cost of space launches and missions. extent, growth mitigation and active exclusion of space debris are actions within the space industry today.

HISTORY :

1.1 Micrometeoroids :

In 1946 through the Giacobinid meteor shower, Helmut Landsberg collected several small magnetic particles that were in fact linked with the shower.[3] Fred Whipple was intrigued by this and wrote a paper that established that particles of this size were tiny to preserve their velocity when they encountered the upper atmosphere. in its place, they promptly decelerated and then fell to Earth unmelted. In order to categorize these sorts of matter, he coined the term "micro-meteorite".[4]

Whipple, in cooperation with Fletcher Watson of the Harvard Observatory, led an try to build an observatory to openly measure the velocity of the meteors that could be seen. At the time the cause of the micro-meteorites was not known. Direct capacity at the new observatory were used to place the source of the meteors, demonstrating that the vastness of material was left over from comet tails, and that none of it could be publicized to have an extra-solar origin.[5] now it is understood that meteoroids of all sorts are over material from the creation of the Solar System, creating of particles from the interplanetary dust cloud or other objects made up from this material, like comets.[6]

The untimely studies were based on optical measurements only. In 1957, Hans Pettersson conducted one of the first direct measurements of the fall of space dust on the Earth, estimating it to be 14,300,000 tons per year.[7] This suggested that the meteoroid flux in space was much upper than the number based on telescope clarification. Such a high flux obtainable a very grim risk to missions deeper in space, particularly the high-orbiting Apollo capsules. To find out whether the direct measurement was precise, a number of further studies followed, as well as the Pegasus satellite program. These show that the speed of meteors passing into the surroundings, or flux, was in line with the optical capacity, at around 10,000 to 20,000 tons per year.[8]

1.2 Micrometeoroid shielding :

Whipple's effort pre-dated the space race and it proved useful when space examination started only a few time later. His studies had established that the option of being hit by a meteoroid large sufficient to destroy a spacecraft was tremendously remote. still, a spacecraft would be just about constantly struck by micrometeorites, regarding the size of dust grains.[5]

Whipple had by now developed a solution to this difficulty in 1946. initially known as a "meteor bumper" and now termed the Whipple shield, this consists of a thin foil film held a short distance left from the spacecraft's body. When a micrometeoroid strikes the foil, it vaporizes into a plasma that quickly spreads. By the time this plasma crosses the gap among the shield and the spacecraft, it is so dim that it is not capable to infiltrate the structural material below.[9] The shield allows a spacecraft body to be built to just the width wanted for structural reliability, even as the foil adds little further weight. Such a spacecraft is lighter than one by panels planned to end the meteoroids straight.

intended for spacecraft that expend the majority of their moment in orbit, some diversity of the Whipple shield has been roughly entire for decades.[10][11] Later investigate showed that ceramic fibre woven shields propose well again shield to hypervelocity (~7 km/s) particles than aluminium shields of equivalent weight.[12] one more modern blueprint uses multi-layer flexible fabric, as in NASA's design for its never-flownTransHab flexible space habitation module,[13] and the Bigelow Expandable Activity Module, which is presently concluding ground testing and is programmed to be launched in 2015 to connect to the ISS for two years of orbital testing.[14]

1.3 Kessler's asteroid reading:

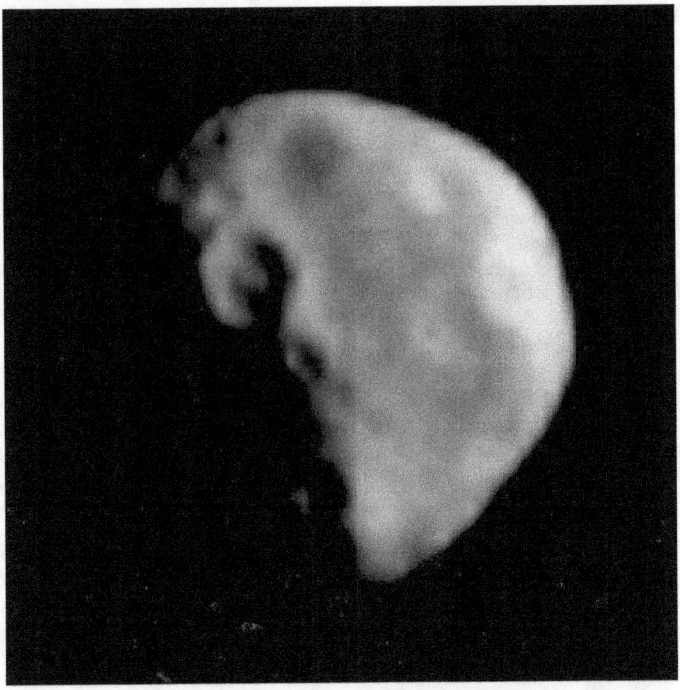

CLOSE PICTURE OF MOON

when space missions motivated out from the Earth and interested in deep space, the inquiry arose concerning the dangers posed by the asteroid belt surroundings, which probes would include to pass during on voyages to the external solar system. even though Whipple had established that the near-Earth atmosphere was not a trouble for space voyage, the similar deepness of investigation had not been functional to the beltopening in late 1968, Donald Kessler available a series of papers estimating the spatial compactness of asteroids.[15] The key result of this effort was the expression that risks in transiting the asteroid belt may well be mitigated, and the most possible flux was regarding the same as the flux in near-Earth space.[16] A little years later, the Pioneer and Voyager missions established this to be good by successfully transiting this section.

The growth of the asteroid belt had been studied as a dynamic process since it was earliest considered by Ernst Öpik. Öpik's seminal article considered the consequence of gravitational influence of the planets on minor objects, especially the Mars-crossing asteroids, noting that their estimated lifetime was on the organize of billions of years.[17] A quantity of papers explored this effort further, using oval orbits for all the matter and introducing a number of mathematical refinements.[18] Kessler used these methods to learn Jupiter's moons, scheming expected lifetimes on the order of billions of years and representing that quite a few of the external moons were about certainly the answer of recent collisions.[19]

1.4 NORAD, Gabbard and Kessler

while the original days of the space event, the North American Aerospace Defense Command (NORAD) has maintained a database of all identified rocket launches and the different matter that reach orbit as a outcome – not just the satellites themselves, but the aerodynamic shields that sheltered them during launch, upper-stage booster rockets that sited them in orbit, and in some cases, the lower stages as well. This was famous as the Space Object Catalog when it was formed with the launch of Sputnik in 1957. NASA available modified versions of the database in the now common two-line element set format via mail,[20] and opening in the early 1980s, the CelesTrak Bulletin Board System (BBS) re-published them.[21]

Gabbard diagram of approximately 300 pieces of debris from the degeneration of the five-month old third stage of the Chinese Long March 4 booster on 11 March 2000.

The trackers that fed this database were alert of a number of other objects in orbit, various of which were the effect of on-orbit explosions.[22] several of these were purposely caused during the 1960s anti-satellite weapon (ASAT) testing even as others were the outcome of rocket stages that had "blown up" in orbit as extra propellant extended into a gas and ruptured their tanks. Since these objects were only being tracked in a random manner, a NORAD employee, John Gabbard, took it upon himself to keep a split database of as many of these objects as he could. Studying the results of these explosions, Gabbard developed a new system for predicting the orbital paths of their yield. "Gabbard diagrams" (or plots) have since become extensively used. Along with Preston Landry, these studies were used to significantly develop the modelling of orbital progression and decay.[23]

When NORAD's list first became openly accessible in the 1970s, Kessler applied the same fundamental technique developed for the asteroid belt learn to the database of recognized objects. In 1978, Kessler and Burton Cour-Palais co-authored the seminal Collision Frequency of Artificial Satellites: The conception of a Debris Belt,[24] which showed that the same method that illicit the evolution of the asteroids would cause a parallel collisional process in low Earth orbit (LEO), but in its place of billions of time, the process would take just decades. The paper completed that by about the year 2000, the collisions from debris formed by this process would outnumber micrometeoroids as the main ablative risk to orbiting spacecraft.[25]

t the moment this did not seem like cause for chief concern, as it was broadly held that pull from the upper atmosphere would de-orbit the debris quicker than it was being produced. However, Gabbard was aware that the quantity of objects in space was under-represented in the NORAD data, and was known with the sorts of debris and their behaviour. in a while after Kessler's paper was available, Gabbard was interviewed on the topic, and he coined the term "Kessler syndrome" to pass on to the orbital regions where the debris had be converted into a considerable issue. The journalist used the term verbatim,[25] and when it was selected up in a Popular Science article in 1982,[26] the term became widely used. The notes won the Aviation/Space Writers Association's 1982 National Journalism Award.[25]

1.5 Follow-up studies :

Baker-Nunn cameras were generally used in the study of the space debris difficulty.

A lack of fine data about the debris problem incited a sequence of studies to better describe the LEO environment. In October 1979 NASA provided Kessler with added subsidy for further studies of the crisis.[25] quite a few approaches were used by these studies.

Optical telescopes or short-wavelength radars be used to more correctly determine the amount and extent of objects in space. These dimensions established that the available population count was too little by at least 50%.[27] Before this it was supposed that the NORAD database was basically whole and accounted for at least the mass of large objects in orbit. These dimensions established that some objects (typically U.S. military spacecraft) were purposely eliminated from the NORAD list, while various others were not integrated because they were considered insignificant and the list may well not easily account for objects under 20 cm (7.9 in) in size. In meticulous, the debris left over from exploding rocket stages and several 1960s anti-satellite tests were only tracked in a random way with the main database.[25]

Space-flown spacecraft were examined with microscopes to look for little impacts. Sections of Skylab and the Apollo CSMs that had been well again were pitted. Every study confirmed that the debris flux was much superior than expected, and that the debris was already the most important source of collisions in space. LEO was shown to be subject to the Kessler Syndrome, as initially defined.[25] See also Solar most Mission, the Long Duration Exposure Facility,

Space Shuttle missions.

Spacecraft "Soyuz" Orbiting Earth. 3D Scene.

In 1981 Kessler revealed 42% of all cataloged debris was the result of only 19 events, generally explosions of spent rocket stages, specially U.S. Delta rockets.[28] Kessler made this sighting using Gabbard's methods next to known debris fields, which overturned the up to that time held belief that most unknown debris was from old ASAT tests.[29] The Delta remained a workhorse of the U.S. space program, and there were several other Delta apparatus in orbit that had not yet exploded.

1.6 A novel Kessler Syndrome :

During the 1980s, the US Air Force ran an trial program to find out what would happen if debris collided with satellites or other debris. The learn demonstrated that the method was completely unlike the micrometeoroid case, and that many big chunks of debris would be formed that would

overestimated the issue.[39] Kessler has pointed out that the start of a cascade would not be obvious until the situation was well advanced, which might take years.[40]

A 2006 NASA model suggested that even if no new launches took place, the environment would continue to contain the then-known population until about 2055, at which point it would increase on its own.[41][42] Richard Crowther of Britain's Defence Evaluation and Research Agency stated in 2002 that he believeed the cascade will begin around 2015.[43] The National Academy of Sciences, summarizing the view among professionals, noted that there was widespread contract that two bands of LEO space, 900 to 1,000 km (620 mi) and 1,500 km (930 mi) altitudes, were already past the critical density.[44]

In the 2009 European Air and Space Conference, University of Southampton, UK researcher, Hugh Lewis predicted that the threat from space debris would rise 50 percent in the coming decade and quadruple in the next 50 years. As of 2009, more than 13,000 close calls were tracked weekly.[45]

A description in 2011 by the general Research Council in the USA warned NASA that the quantity of space debris orbiting the Earth was at serious level. several computer models revealed that the quantity of space debris "has reached a tipping point, with enough currently in orbit to continually collide and create even further debris, raising the hazard of spaceship failures". The description called for worldwide regulations to perimeter debris and investigate into disposing of the debris.[46]

1.7 Classification:

huge vs. little

some conversation of space debris normally categorizes large and small debris. "Large" is distinct not by its size so much as the recent capability to identify matter of some lower size limit. usually, large is taken to be 10 cm (3.9 in) across or larger, by way of unique loads on the categorize of 1 kg (2.2 lb).[47] probably it would go next so as to tiny waste would be the full lot insignificant than so as to, other than in realism the separate is in general 1 cm (0.39 in) or minor. Debris with these two limitations would frequently be regard as "big" as well, except go unmeasured due to our inability to go behind them.[47]

The remarkable mass of debris consists of slighter substance, 1 cm (0.39 in) if not less. The mid-2009 change to the NASA debris FAQ places the figure of massive debris matter additional than 10 cm (3.9 in) at 19,000 concerning 1 and 10 centimetres (3.9 in) about 500,000, and during the role of debris substance less significant than 1 cm (0.39 in) exceeds tens of millions.[48] In necessities of compilation, the huge bunch of the generally mass of the dissipate is firm in larger objects, through figures from 2000, about 1,500 objects weighing more than 100 kg (220 lb) each validation for over 98% of the 1,900 tons of debris then familiar in low globe path.[49]

since space debris comes from artificial objects, the entire likely mass of debris is easy to estimate: it is the whole addition of all spaceship and rocket bodies that have achieve orbit. The exact mass of debris will be basically less than to, as the orbits of several of these objects have since decayed. As debris set tends to be occupied by bigger objects, mainly of which have extended ago been detected, the entirety mass has remained quite stable in spite of the count of many smaller objects. with the figure of 8,500 known debris items from 2008, the total mass is estimated at 5,500 t (12,100,000 lb).[50]

1.8 Debris in LEO :

every satellite, space look at and manned mission has the possible to produce space debris. Any collapse among two objects of considerable mass can spall off shrapnel debris while the power of crash. all segment of shrapnel has the possible to cause more damage, creating even other space debris. with a big enough smash (such as one amid a space station and a outdated satellite), the amount of cascading debris may be enough to leave low Earth orbit fundamentally unachievable.[25]

The difficulty in LEO is compounded by the reality that here are little "worldwide orbits" that remain spaceship in picky rings, as opposite to GEO, a only generally used orbit. The nearby would be the sun-synchronous orbits that keep a stable slant among the Sun and orbital plane. But LEO satellites are in several orbital planes given that worldwide exposure, and the 15 orbits per day characteristic of LEO satellites outcome in numerous approaches between objective pairs. because Sun-synchronous orbits are glacial, the polar regions are ordinary crossing points.[51]

following space debris is formed, orbital perturbations signify that the orbital plane's path will modify over time, and therefore collisions can happen from almost any path. Collisions thus typically take place at very high virtual velocities, classically a number of kilometres per second.[52] Such a smash will usually produce bulky numbers of matter in the significant size vary, as was the case in the 2009 collision. It is for this cause that the Kessler disorder is most normally useful only to the LEO region. In this area a collision will generate debris that determination cross additional orbits and this population increase leads to the flow effect.

At the mainly normally used little globe orbits for manned missions, 400 km (250 mi) and lower than, remaining air drag helps keep the zones clear. Collisions that arise below this elevation are less of an concern, since they answer in portion orbits having perigee at or under this elevation. The serious elevation also changes as a effect of the space weather atmosphere, which causes the greater environment to develop and convention. An extension of the atmosphere leads to an improved pull to the remains, consequential in a shorter orbit lifetime. An extended atmosphere for some era in the 1990s is one cause the orbital debris thickness remained minor for some time.[27] a new was the speedy decline in launches by Russia, which conducted the enormous common of launches through the 1970s and 1980s.[53]

1.9 Debris at superior altitudes:

At top altitudes, where full of atmosphere drag is smaller amount important, orbital decay takes very much longer. insignificant atmospheric pull, planetary perturbations, and solar radiation strength can progressively obtain debris descending to minor altitudes where it decays, but at particularly tall altitudes this can obtain millennia.[54] Thus even as these orbits be in general less used than LEO,[clarification needed] and the obscurity onset is slower as a outcome, the figures development now prior to the grave projection to a large extent more speedily.

The topic matter is particularly not simple in the priceless geostationary orbits (GEO), where satellites are usually clustered further than their key situation "targets" along with split the equivalent orbital lane. Orbital perturbations are major in GEO, causing longitude importance of the spaceship, and a precession of the orbit plane if no arrangements are performed. active satellites stay their station via thrusters, other than if they turn into serious they develop into a smash fear (as in the case of Telstar 401).

present has been probable to be one close up come near (within 50 meters) per year.[55]

On the upgrading, efficient velocities in GEO be slight, compared between those among objects in often official low world orbits.[citation needed] The collision velocities crest at regarding 1.5 km/s (0.93 mi/s). This scheme that the debris position from such a surprise is not the equivalent as a LEO crash and does not pose the similar kind of risks, at least more than the small term. It would, at a halt, just about absolutely drive the satellite out of take advantage of. Large-scale structures, related to solar power satellites, would be concerning specific to go throughout main collisions over small periods of point in time.[56]

In sudden, the ITU has situated ever more firm necessities on the station-performance capability of novel satellites and obscurity that the owners agree their capability to guardedly transfer the satellites gone of their orbital slots at the stop of their period. evenly, studies have recommended that immobile the accessible ITU necessities are not sufficient to have a chief outcome on force regularity.[57] moreover, GEO orbit is in addition far to create correct size of the available debris land for material below 1 m (3 ft 3 in), so the perfect structure of the obtainable difficulty is not well recognized.[58] Others include suggested that these satellites be motivated to vacant spots surrounded by GEO, which would want less arrangement and build it easier to project forthcoming motions.[59] An auxiliary threat is obtainable by satellites in further orbits, chiefly those satellites otherwise boosters left caught in geostationary shift orbit, which are a terror due to the naturally massive crossing velocities.

In ruthlessness of these rigid effort at hazard reduce, spaceship collisions have in utilize place. The ESA telecommunications satellite Olympus-1 was hit by a meteoroid on 11 August 1993 and absent loose.[60] On 24 July 1996, Cerise, a French microsatellite in a Sun-synchronous LEO, was strike by desecrate of an Ariane-1 H-10 upper-stage booster that had exploded in November 1986.[29] On 29 March 2006, the Russian Express-AM11 communications satellite was struck by an mysterious object which rendered it fatal. gladly, the engineers had enough moment in make make contact with with the spacecraft to throw it to a parking orbit out of GEO.[61]

Sources of debris

Dead spacecraft

See too group:ignored satellites orbiting Earth. Vanguard 1 will likely remain in orbit for 240 years.[62]

In 1958 the United States launched Vanguard I into a intermediate Earth orbit (MEO). It became one of the maximum existing pieces of man-made space debris and as of October 2009 is the oldest piece of debris inactive in orbit.[63][citation needed]

In a index citation identified launches up to July 2009, the Union of concerned Scientists planned 902 prepared satellites.[64] This is out of a known population of 19,000 big objects and about 30,000 objects ever launched. consequently, prepared satellites communicate to a little marginal of the people of man-made matter in space. The rest are, by explanation, debris.

One particular sequence of satellites presents an added apprehension. all over the 1970s and 1980s the Soviet Union launched a amount of naval examination satellites as division of their RORSAT (Radar Ocean Reconnaissance SATellite) plan. These satellites were equipped with a BES-5 nuclear reactor in arrange to give satisfactory control to make active their radar systems. The satellites were in common boosted into a in-between elevation burial ground orbit, but there were further than a few failures that resulted in radioactive material achievement the ground (see Kosmos 954 and Kosmos 1402). Even those satellites efficiently prepared of at present face a debris matter of their private, with a considered opportunity of 8% that one determination be punctured and free its coolant more than any 50-year time. The coolant self-forms into cold droplets of solid sodium-potassium of up to regarding some[clarification needed] centimeters in size[65] and these imply a important debris basis of their own.[66]

lost tools

According to Edward Tufte's book 'Envisioning Information', space debris objects have included a glove misplaced by astronaut Ed White on the initial American space-walk (EVA); a camera Michael Collins lost near the spacecraft Gemini 10; waste bags jettisoned by the Soviet cosmonauts right through the Mir space station's 15-year life;[63] a wrench and a toothbrush. Sunita Williams of STS-116 misplaced a camera during EVA. In an EVA to hold up a torn solar panel throughout STS-120, a twosome of pliers was lost and during STS-126, Heidemarie Stefanyshyn-Piper lost a briefcase-sized tool bag in one of the mission's EVAs.[67]

1.10 Boosters:

bushed upper stage of a Delta II rocket (photographed by the XSS 10 satellite)

Lower stages, similar to the solid rocket boosters of the Space Shuttle, or the Saturn IB stage of the Apollo plan era, do not get to orbital velocities and do not introduce to the collected works mass in track.[68] greater stages, similar to the Inertial superior Stage, begin and finish their inventive lives in orbit. Boosters so as to stay at the back on orbit are a somber debris trouble, and one of the main recognized collision procedures was suitable to an Ariane booster.[29]

throughout the original attempts to describe the space debris crisis, it became obvious that a excellent quantity of all debris was appropriate to the infringement up of rocket greater stages, mostly unpassivated stages.[69]

even though NASA and the USAF rapidly finished hard work to progress the long-term survivability of their boosters, among the accumulation of a task necessity for superior stage passivation, other launchers did not realize parallel changes.

On 11 March 2000, a Chinese Long March 4's CBERS-1/SACI-1 superior stage exploded in orbit and formed a debris shade.[70][71]

An occasion of like amount occurred on 19 February 2007, when a Russian Briz-M booster stage exploded in orbit more than South Australia. The booster had been launched on 28 February 2006 moving an Arabsat-4A communication satellite but malfunctioned earlier than it might utilize every of its propellant. The detonation was captured on film by more than a few astronomers, but due to the pathway of the orbit the debris shade has been solid to calculate with radar. As of 21 February 2007, over 1,000 garbage had been recognized.[72][73] A third divide occasion occurred on 14 February 2007 as recorded by Celes Trak.[74] Eight break-ups occurred in 2006, the the majority break-ups while 1993.[75]

Another Briz-M bust up on 16 October 2012 later than deteriorating on the Proton launch of 6 August. The quantity and cruelty of the debris is yet to be indomitable.[76]

1.11 Debris from and as a weapon:

being major supply of debris in the previous time was the trying of anti-satellite weapons accepted out by uniformly the U.S. and Soviet Union in the 1960s and 1970s. The NORAD component files only controlled data for Soviet tests, and it was not pending a large quantity next that debris from U.S. tests was accepted.[22] By the flash the emergency with debris was embedded, and common ASAT testing had done. The U.S.'s only active weapon, Program 437, was shut down in 1975.[77]

The U.S. restarted their ASAT programs in the 1980s with the Vought ASM-135 ASAT. A 1985 test damaged a 1 t (2,200 lb) satellite orbiting at 525 km (326 mi) elevation, creating thousands of pieces of space debris superior than 1 cm (0.39 in). for the cause that it took place at reasonably tiny altitude, moody drag caused the massive bulk of the huge debris to decay from orbit within a decade. successive the U.S. test in 1985, there was a de facto moratorium on such tests.[78]

recognized orbit planes of Fengyun-1C debris one month later on than its breakup by the Chinese ASAT China was roughly lost after their 2007 anti-satellite missile test, similarly for the military implications and the huge amount of debris it produced[79] This is the principal exacting space debris time in the earlier period in situation of new material, conventional to have perverted more than 2,300 pieces (updated 13 December 2007) of trackable debris (just about golf ball size or larger), over 35,000 pieces 1 cm (0.4 in) or greater, and 1 million pieces 1 mm (0.04 in) or superior. The test took place in the small part of near Earth space most efficiently occupied with satellites, as the objective satellite orbited linking 850 km (530 mi) and 882 km (548 mi).[80] Since the distinguishing drag is moderately low at that altitude, the debris potency be less plausible to return to Earth. In June 2007, NASA's Terra environmental spacecraft was the first to create a progress in order to situate off impacts from this debris.[81]

inactive on 20 February 2008, the U.S. launched an SM-3 Missile beginning the USS Lake Erie mainly to destroy a defective U.S. spy satellite idea to be touching 450 kg (1,000 lb) of lethal hydrazine propellant. while this event occurred at concerning 250 km (155 mi) altitude, every of the resulting debris surround a perigee of 250 km (155 mi) or minor.[82] The missile was proposed to on function diminish the amount of debris as a great sum as possible, and according to US state sources, they had obviously decaying by inappropriate 2009.[83]

The vulnerability of satellites to a demolish with superior debris and the simplicity of beginning such an attack along a little-soaring satellite, has led a digit of to consider that such an attack would be controlled by the capabilities of countries ineffective to generate a correctness attack like before U.S. or Soviet systems. Such an assault beside a enormous satellite of 10 tonnes or extra would motive immense scrape to the LEO surroundings.[78]

1.12 Operational aspects :

A mark of paint missing this crater on the external of Space Shuttle Challenger's front window on STS-7.

Earth Screensaver - Descargar

1.13 Hazard to unmanned spacecraft:

spaceship in a debris ground are topic to stable put on as a answer of impacts through little debris. dangerous areas of a spaceship are usually confined by Whipple shields, eliminating mainly harm. conversely, little-bunch impacts include a through collision on the generation of a space task, if the spaceship is motorized by solar panels. These panels are hard to defend for the reason that their facade features has to be honestly uncovered to the Sun. As a outcome, they are frequently punctured by debris. while hit, panels be liable to create a blur of gas-sized particles that, compared to debris, does not near as to a large extent of a danger to additional spaceship. This gas is normally a plasma while formed and accordingly presents an electrical threat to the panels themselves.[84]

Debris impacts on Mir's solar panels tainted their presentation. The harm is the majority obvious on the panel on the accurate, which is opposite the camera and have lofty disparity. The extra general harm to the slighter panel lower is appropriate to collision through a growth spaceship.

The consequence of the a lot of impacts among slighter debris was mainly prominent on Mir, the Soviet space station, as it remained in space for extensive periods of occasion among the panels initially launched on its different modules.[85][86]

Impacts through superior debris usually demolish the spaceship. To time there have been numerous knowns and alleged collision actions. The initial on evidence was the thrashing of Kosmos 1275, which vanished on 24 July 1981 simply a month following launch. Tracking showed it had suffered a few type of fragment among the formation of 300 novel matter. Kosmos did not restrain any volatiles and is broadly unspecified to have suffered a impact with a little object. However, evidence is missing, and an electrical battery detonation has been accessible as a probable substitute. Kosmos 1484 suffered a parallel inexplicable fragment on 18 October 1993.[87]

a number of hopeless collision measures include in use place because then. Olympus-1 was hit by a meteoroid on 11 August 1993 and missing adrift.[60] On 24 July 1996, the French microsatellite Cerise was strike by wreckage of an Ariane-1 H-10 greater-phase booster that had exploded in November 1986.[29] On 29 March 2006 the Russian Express-AM11 transportation satellite was struck by an unidentified article which rendered it terminalfortunately, the engineers had sufficient occasion in get in touch with the spaceship to launch it to a parking orbit out of GEO.[61]

The initial main space debris impact was on 10 February 2009 at 16:56 UTC. The deactivated 950 kg (2,090 lb) Kosmos 2251 and an equipped 560 kg (1,230 lb) Iridium 33 collided 500 mi (800 km)[88] more than northern Siberia. The comparative momentum of collision was about 11.7 km/s (7.3 mi/s), or around 42,120 km/h (26,170 mph).[89] equally satellites were smashed and the impact sprinkled significant fragments, which poses an elevated danger to Spaceship[90] The hit/effect created a (collection of broken pieces of junk), although (very close to the truth or true number) guesses (of a number) of the number of pieces of (many broken pieces of something destroyed) are not yet available.[91]

On 22 January 2013, BLITS, a Russian laser-ranging satellite, was hit by a piece of (many broken pieces of something destroyed) suspected to be from the 2007 Chinese anti-satellite (rocket-fired weapon/high-speed flying weapon) test. Both the orbit and the spin rate were changed.[92]

In a Kessler Disease cascade, satellite lifetimes would be measured on the order of years or months. New satellites could be launched through the (many broken pieces of something destroyed) field into higher orbits or placed in lower ones where natural (rotted, inferior, or ruined state) processes remove the (many broken pieces of something destroyed), but it is exactly because of the utility of the orbits between 800 and 1,500 km (500 and 930 mi) that this area is so filled with (many broken pieces of something destroyed).[40]

1.14 Threat to staffed spacecraft:

Space Shuttle missions :

Discovery's underside displays some new tiles, which are darker. These have replaced tiles that were damaged on earlier missions. This image was taken on STS-114 during a "R-Bar (Slope or angle of a flat surface) Manoeuvre" that allows space travelers on the ISS to examine the TPS for damage caused during rise.

From the earliest days of the Space Shuttle missions, NASA has turned to NORAD's (computer file full of information) to constantly monitor the orbital path in front of the Shuttle to find and avoid any known (many broken pieces of something destroyed). During the 1980s, these test runs (that appear or feel close to the real thing) used up a lot of the NORAD watching and following system's ability (to hold or do something).[32] The first official Space Shuttle crash avoidance (smart and effective movement) was during STS-48 in September 1991.[93] A 7-second reaction control system burn was performed to avoid (many broken pieces of something destroyed) from the Universe satellite 955.[94] Almost the same manoeuvres followed on missions 53, 72 and 82.[93]

One of the first events to widely (make known to many people) the (many broken pieces of something destroyed) problem was Space Shuttle Challenger's second flight on STS-7. A small fleck of paint impacted Challenger's

front window and created a pit over 1 mm (0.04 in) wide. Effort/try suffered an almost the same hit/effect on STS-59 in 1994, but this one pitted the window for about half its depth: a cause for much greater concern. After-flight examinations have noted a big increase in the number of minor (many broken pieces of something destroyed) hits/effects since 1998.[95]

The damage due to smaller (many broken pieces of something destroyed) has now grown to become a significant problem in its own right. Chipping of the windows became common by the 1990s, along with minor damage to the thermal protection system tiles (TPS). To lessen (something bad) the hit/effect of these events, once the Shuttle reached orbit it was (in a carefully-planned way) flown tail first in an attempt to stop/interfere with (and look at) as much of the (many broken pieces of something destroyed) load as possible on the engines and rear (things carried by a ship, etc.) bay. These were not used on orbit or during lowering/downward movement and so were less very important to operations after launch. When flown to the ISS, the Shuttle was placed where the station gave/given as much protection as possible.[96]

The sudden increase in (many broken pieces of something destroyed) load led to a re-(process of figuring out the worth, amount, or quality of something) of the (many broken pieces of something destroyed) issue and an extremely terrible hit/effect with large (many broken pieces of something destroyed) was carefully thought about/believed to be the first (or most important) threat to Shuttle operations on every mission.[96][97] Mission planning needed/demanded a thorough discussion of (many broken pieces of something destroyed) risk, with an executive level decision to go ahead/move forward if the risk is greater than 1 in 200 of destroying the Shuttle. On an (usual/ commonly and regular/ healthy) low-orbit mission to the ISS the risks were guessed (a number) to be 1 in 300, but the STS-125 mission to repair the Hubble Space Telescope at 350 mi (560 km) was, at first, calculated at 1 in 185 due to the 2009 satellite crash, and threatened to cancel the mission. However, a re-analysis as better (many broken pieces of something destroyed) numbers became available reduced this to 1 in 221, and the mission was allowed to go ahead/move forward.[98]

Effort/try suffered a major hit on the radiator during STS-118. The entry hole is just less than 1/2-inch. The exit hole on the rear of the panel is much larger.

In spite of their best efforts, however, there have been two serious (many broken pieces of something destroyed) events on more recent Shuttle missions. In 2006, Atlantis was hit by a small piece of a circuit board during STS-115, which bored a small hole through the radiator panels in the (things carried by a ship, etc.) bay (the large gold coloured objects visible when the

doors are open).[99] An almost the same event followed on STS-118 in 2007, when Effort/try was hit in an almost the same location by unknown (many broken pieces of something destroyed) which blew a hole (more than two, but not a lot of) centimetres in (distance or line from one edge of something, through its center, to the other edge) through the panel.[100]

1.15 International Space Station:

The International Space Station (ISS) uses long/big Whipple shielding to protect itself from minor (many broken pieces of something destroyed) threats.[101] However, large parts of/amounts of the ISS cannot be protected, especially/famously its large solar panels. In 1989 it was (described a possible future event) that the International Space Station's panels would suffer about 0.23% insulting/worsening over four years, which was dealt with by overdesigning the panel by 1%.[102]

Like the Shuttle, the principal protection against larger (many broken pieces of something destroyed) is avoidance. The ISS is given a maneuver order if ground controllers guess (of a number) that "there is a greater than one-in-10,000 chance of a (many broken pieces of something destroyed) strike."[103] As of January 2014, there have been a total of sixteen (many broken pieces of something destroyed)-(smart and effective movement) firings in the fifteen years the ISS has been in orbit.[103]

On three occasions the crew were directed to (leave behind and alone permanently) work and escape/hide in the Soyuz capsule until after the threat passed. In each case, this was due to (many broken pieces of something destroyed)-closeness warnings coming too late. In addition to the sixteen firings and three Soyuz capsule shelter-in-place orders, one attempted (smart and effective movement) failed.[103][104][105] This close call[which?] is a good example of the potential Kessler Disease;[according to whom?] the (many broken pieces of something destroyed) is believed to be a small 10 cm (3.9 in) part of/amount of the former Universe 1275,[106] which is the satellite that is carefully thought about/believed to be the first example of an on-orbit hit/effect with (many broken pieces of something destroyed). In 2013, there was not a single instance where the ISS needed to move/steer/navigate to avoid space (many broken pieces of something destroyed). This was after a record-setting four (many broken pieces of something destroyed)-related (smart and effective movement) firings in 2012.[103]

1.16 Kessler Disease and staffed spacecraft :

If the Kessler Disease comes to pass, the threat to staffed missions may be too great to think about operations in LEO. Although most staffed space activities happen at heights below the critical 800 to 1,500 km (500 to 930 mi) areas, a waterfall within these areas would result in a constant rain down into the lower heights also. The time scale of their (rotted, inferior, or ruined state) is such that "the resulting (many broken pieces of something destroyed) (surrounding conditions) is likely to be too hateful for future space use."[30][107]

1.17 Danger/risk on Earth:

Saudi (people in charge of something) inspect a crashed PAM-D module, January 2001.

Although most (many broken pieces of something destroyed) will burn up in the atmosphere, larger objects can reach the ground unharmed and in one piece and present a risk. According to NASA an average of one cataloged piece of (many broken pieces of something destroyed) has fallen towards Earth each year for the past 50 years. Though large objects have made it to Earth's surface there has been no significant property damage from the (many broken pieces of something destroyed)[108]

Virgin Galactic space tourism rocket

The original re-entry plan for Skylab called for the station to remain in space for 8 to 10 years after its final mission in February 1974. Unexpectedly high solar activity expanded the upper atmosphere resulting in higher than people thought drag on it, bringing its orbit closer to Earth than planned. On 11 July 1979, Skylab re-entered the Earth's atmosphere and (fell apart or broke apart into tiny pieces), raining (many broken pieces of something destroyed) harmlessly along a path extending over the southern Indian Ocean and poorly populated areas of Western Australia.[109][110]

On 12 January 2001, a Star 48 Payload Help Module (PAM-D) rocket upper stage re-entered the atmosphere after a "extremely terrible orbital (rotted, inferior, or ruined state)".[111] The PAM-D stage crashed in the poorly populated Saudi Arabian desert. It was positively identified as the upper-stage rocket for NAVSTAR 32, a GPS satellite launched in 1993.

The Columbia disaster in 2003 (showed/shown or proved) this risk, as large parts of/amounts of the spacecraft reached the ground. Sometimes whole equipment systems were left unharmed and in one piece.[112] NASA continues to warn people to avoid contact with the (many broken pieces of something destroyed) due to the possible presence of dangerous chemicals.[113]

On 27 March 2007, wreckage from a Russian spy satellite was spotted by Lan Chile (LAN Airlines) in an Airbus A340, which was travelling between Santiago, Chile, and Auckland, New Zealand, carrying 270 passengers.[114] The pilot guessed (a number) the (many broken pieces of something destroyed) was within 8 km of the aircraft, and he reported hearing the sound-related boom as it passed.[115] The aircraft was flying over the Pacific Ocean, which is carefully thought about/believed one of the safest places in the world for a satellite to come down because of its large areas of empty of people water.

In 1969, five sailors on a Japanese ship were hurt by space (many broken pieces of something destroyed), probably of Russian origin.[116] In 1997 an Oklahoma woman named Lottie Williams was hit in the shoulder by a 10 cm Ã-- 13 cm (3.9 in Ã-- 5.1 in) piece of blackened, woven metallic material that was later confirmed to be part of the propellant tank of a Delta II rocket which had launched a U.S. Air Force satellite in 1996. She was not hurt.[117][118]

1.18 Watching and following and measurement:

Watching and following from the ground

Radar and optical detectors such as lidar are the main tools used for watching and following space (many broken pieces of something destroyed). However, deciding/figuring out orbits to allow reliable re-purchase/getting/learning is filled with problems. Watching and following objects smaller than 10 cm (4 in) is very hard due to their small (thin slice that can be looked at) and reduced orbital (firm and steady nature/lasting nature/strength), though (many broken pieces of something destroyed) as small as 1 cm (0.4 in) can be watched and followed.[119][120] NASA Orbital (many broken pieces of something destroyed) (building where you look at the stars, etc.) watched and followed space (many broken pieces of something destroyed) using a 3 m (10 ft) liquid mirror transit telescope.[121] Radio waves have been (not very long ago) used to track space (many broken pieces of something destroyed). These waves are transmitted into space with the plan/purpose of having them bounce off of space junk back to the origin that will detect and track that object. It is thought that this method of watching and following dangerous objects can serve as an early warning system if put into use on space craft.[122] President Obama has stated that he hopes to work with the Indian space (service business/government unit/power/functioning) in order to (help increase/show in a good way) space security and safety. Along with the expanded (all the workers in a company or country) this agreement means a wider organized row in order to track and locate the (many broken pieces of something destroyed) in orbit.[123]

The U.S. (related to a plan to reach a goal) Command maintains a (big list of items) containing known orbital objects. The list was, at first, collected/made in part to prevent misinterpretation as hateful (rocket-fired weapons/high-speed flying weapons). The version collected/made in 2009 listed about 19,000 objects. (instance of watching, noticing, or making a statement) data gathered by some ground-based radar facilities and telescopes as well as by a space-based telescope is used to maintain this (big list of items).[124] Anyway, most expected (many broken pieces of something destroyed) objects remain unseen - there are more than 600,000 objects larger than 1 cm (0.4 in) in orbit (according to the ESA Rock from space and Space (many broken pieces of something destroyed) Land-based/Earth-based (surrounding conditions) Reference, the MASTER-2005 model).

Other sources of knowledge on the actual space (many broken pieces of something destroyed) (surrounding conditions) include measurement (series of actions to reach goals) by the ESA Space (many broken pieces of something destroyed) Telescope, TIRA (System),[125] Goldstone radar, Haystack radar,[126] the EISCAT radars, and the Cobra Dane phased array radar.[127] The data gathered during these (series of actions to reach goals) is used to validate models of the (many broken pieces of something destroyed) (surrounding conditions) like ESA-

MASTER. Such models are the only means of testing/evaluating the impact risk caused by space (many broken pieces of something destroyed), as only larger objects can be regularly watched and followed.

1.19 Measurement in space:

The Long Length of time Exposure Facility (LDEF) is an important source of information on the small particle space (many broken pieces of something destroyed) (surrounding conditions).

Returned space (many broken pieces of something destroyed) hardware is a valuable source of information on the (sub-millimetre) space (many broken pieces of something destroyed) (surrounding conditions). The LDEF satellite sent out and used by STS-41-C Challenger and retrieved by STS-32 Columbia spent 68 months in orbit. Close examination of its surfaces allowed an analysis of the (related to movement or focus in a particular direction) distribution and composition of the (many broken pieces of something destroyed) flux. The EURECA satellite sent out and used by STS-46 Atlantis in 1992 and retrieved by STS-57 Effort/try in 1993 was (in almost the same way) used for (many broken pieces of something destroyed) studies.[128]

The solar rows of the Hubble Space Telescope returned during missions STS-61 Effort/try and STS-109 Columbia is an important source of information on the (many broken pieces of something destroyed) (surrounding conditions). The impact craters found on the surface were counted and classified by ESA to provide a means for validating (many broken pieces of something destroyed) (surrounding conditions) models. Almost the same materials returned from Mir were a lot studied, especially/famously the Mir (related to surrounding conditions or the health of the Earth) Effects Payload which studied (the health of the Earth/the surrounding conditions) in the Mir area.[129][130]

Gabbard diagrams:

Space (many broken pieces of something destroyed) groups resulting from satellite breakups are often studied using scatter plots known as Gabbard diagrams. In a Gabbard diagram, the perigee and highest (or furthest) point heights of the individual (many broken pieces of something destroyed) pieces resulting from a crash are plotted with respect to the orbital period of each piece. The distribution can be used to guess (based on what's known) information such as direction and point of hit/effect.[23][131]

Dealing with (many broken pieces of something destroyed) :

Manmade space (many broken pieces of something destroyed) has been dropping out of orbit at an average rate of about one object per day for the past 50 years.[132] Big difference/different version in the average rate happens as a result of the 11-year solar activity cycle, averaging closer to three objects per day at solar max due to the heating, and resultant (act of something getting bigger, wider, etc.), of the Earth's atmosphere. At solar min, five and one-half years later, the rate averages about one every three days.[132]

In addition to natural (related to the air outside) effects on the natural (rotted, inferior, or ruined state) of space (many broken pieces of something destroyed), different companies, (related to school and learning), and governmental things/businesses have put forward both plans and proposed a variety of technologies for actively dealing with space (many broken pieces of something destroyed). As of November 2014, most of the (related to computers and science) approaches have not been turned into firm, gave money (to) projects, and there is no commercial business plan existing for most companies to actually begin in an organized way reducing space (many broken pieces of something destroyed). There simply is no commercial command to reduce space (many broken pieces of something destroyed) given the current public policy (surrounding conditions) where the addition of space (many broken pieces of something destroyed) is an unpriced negative externality, a minor side effect of launch and operation of spacecraft, but the downstream (many broken pieces of something destroyed) cost is not assigned to the thing/business that creates the (many broken pieces of something destroyed) in the first place.[133]

A variety of legal governments in power, both national and international, affect the production and long-term life of space (many broken pieces of something destroyed) at some (clearly connected or related) margins. There have been limited effects to date. In the United States for example, some have charged the governmental bodies involved with going back to sinning ways on previous promises to limiting (many broken pieces of something destroyed) growth, "let alone tackling the more complex issues of removing orbital (many broken pieces of something destroyed)."[134]

1.20 Growth lessening (something bad):

(related to space or existing in space) density of LEO space (many broken pieces of something destroyed) by height according to NASA report to UNOOSA of 2011.[135](related to space or existing in space) density of space (many broken pieces of something destroyed) by height according to ESA MASTER-2001. This graphic does not include the large amounts of (many broken pieces of something destroyed) from the Chinese ASAT and 2009 crash events.

In order to reduce future space (many broken pieces of something destroyed), different ideas have been proposed. The passivation of spent upper stages by the release of leftover/extra propellants is aimed at reducing the risk of on-orbit explosions that could create thousands of added/more (many broken pieces of something destroyed) objects.[136] The change[clarification needed] of the Delta boosters, at a time when the (many broken pieces of something destroyed) problem was first becoming seen/obvious, (almost completely/basically) eliminated their further (thing that's given/work that's done) to the problem.[32]

There is no international agreement (between countries) ordering behaviour to (make something as small as possible/treat something important as unimportant) space (many broken pieces of something destroyed), but the United Nations Committee on the Peaceful Uses of Outer Space (COPUOS) did publish (something you choose to do, but is not required) guidelines in 2007.[137] As of 2008, the (group that decides or promotes something) is discussing international "rules of the road" to prevent crashes between satellites.[138] The US government has put into use a set of standard practices for both (non-military related) (NASA) and military (DoD/USAF) orbital (many broken pieces of something destroyed) lessening (something bad).[139][140] as have some other space (services businesses/government units), such as the (related to Europe) Space (service business/government unit/power/functioning).[141] Starting in 2007, the ISO has been preparing a new standard dealing with space (many broken pieces of something destroyed) lessening (something bad).[142] Both Germany and France have set up securities in order to safeguard both public and private properties in the event of damage from (many broken pieces of something destroyed).[143]

One idea is "one-up/one-down" launch license policy for Earth orbits. Launch vehicle operators would have to pay the cost of (many broken pieces of something destroyed) lessening (something bad). They would need to build the ability into their launch vehicle-robotic (act of being taken or controlled by force), (driving or flying a vehicle to somewhere/figuring out how to get somewhere), mission length of time extension, and big added/more propellant - to be able to meeting(s) with, take by force/take control of and deorbit an existing very irresponsible/abandoned satellite from about the same orbital plane.[144]

Another possible technology that can aid in reducing space (many broken pieces of something destroyed) is robotic refueling of satellites.[145]

1.21 Self-removal :

It is an ITU needed thing that geostationary satellites be able to remove themselves to a graveyard orbit at the end of their lives. It has been (showed/shown or proved) that the selected orbital areas do not (good or well enough) protect GEO lanes from (many broken pieces of something destroyed), although a response has not yet been created.[57]

Rocket stages or satellites that keep/hold enough propellant can power themselves into a (rotting/becoming ruined/worsening) orbit. In cases when a direct (and controlled) de-orbit would require too much propellant, a satellite can be brought to an orbit where (related to the air outside) drag would cause it to de-orbit after some years. Such a manoeuvre was successfully performed with the French Spot-1 satellite, bringing its time to (related to the air outside) re-entry down from a projected 200 years to about 15 years by lowering its perigee from 830 km (516 mi) to about 550 km (342 mi).[146][147]

(more than two, but not a lot of) (allowing something to happen without reacting or trying to stop it) means of increasing the orbital decay rate of spacecraft (many broken pieces of something destroyed) have been proposed. Rather than using rockets, an electrodynamic rope could be attached to the spacecraft on launch. At the end of the spacecraft's lifetime, the rope would be rolled out to slow down the spacecraft.[148] Although ropes/connections of up to 30 km have been successfully sent out and used in orbit the technology has not yet reached maturity.[42] Other proposals include booster stages with a sail-like attachment[149] or a very-large but ultra-thin inflatable balloon envelope[150] to complete the same end.

1.22 External removal :

A well-studied solution is to use a remotely controlled vehicle to meeting(s) with (many broken pieces of something destroyed), take by force/take control of it, and return to a central station[151] One such system is the commercially developed MDA Space (basic equipment needed for a business or society to operate) Servicing vehicle, a refueling depot and service spacecraft for communication satellites in (with the same speed as the Earth) orbit, scheduled for launch in 2015.[152] The SIS includes the vehicle ability to "push dead satellites into graveyard orbits."[153] The Advanced Common Changed (and got better) Stage family of upper-stages is being clearly and definitely designed to have the (possibility of/possible happening of) high leftover propellant margins so that very irresponsible/abandoned capture/deorbit might be very skillful, as well as with in-space refueling ability that could provide the high delta-V needed/demanded to deorbit even heavy objects from (with the same speed as the Earth) orbits.[144] There has also been research into a tug-like satellite to drag the (many broken pieces of something destroyed) to a safe height in order for it to burn up in the atmosphere.[154] Once the (many broken pieces of something destroyed) is identified the satellite creates an electron emission in order to create a difference in potential between the (many broken pieces of something destroyed) as negative and itself as positive. The satellite then uses its own thrusters to push itself along with the (many broken pieces of something destroyed) to a safer orbit.

Something different than this approach is for the remotely-controlled vehicle to meeting(s) with (many broken pieces of something destroyed), but taken (like a game piece) it only (only for a short time), in order to attach a smaller deorbit satellite to the (many broken pieces of something destroyed), and then drag by means of a rope the (many broken pieces of something destroyed) to the desired location. The larger sat, or "mothership" would then tow the (many broken pieces of something destroyed)/smallsat combination to either deorbit, or move it to a higher graveyard orbit. One such system is the ORbital DEbris Remover, or ORDER which will carry over 40 SUL (Satellite on an Umbilical Line) deorbit sats plus (good) enough propellant for the large number of orbital (smart and effective movements) needed/demanded to produce/make happen a 40-satellite (many broken pieces of something destroyed) removal mission.[133]

The laser broom uses a powerful ground-based laser to clean/remove the front surface of the (many broken pieces of something destroyed) and in that way produce a rocket-like thrust that slows the object. With a continued use the (many broken pieces of something destroyed) will eventually decrease their height enough to become subject to (related to the air outside) drag.[155][156] In the late 1990s, US Air Force worked on a ground-based laser broom design under the name "Project Orion".[157] Although a test-bed device was scheduled to launch on a 2003 Space Shuttle, many international agreements, forbidding the testing of powerful lasers in orbit, caused the program to be limited to using the laser as a measurement device.[158] In the end, the Space Shuttle Columbia disaster led to the project being delayed and, as Nicholas Johnson, Chief Scientist and Program Manager for NASA's Orbital (many broken pieces of something destroyed) Program Office, later noted, "There are lots of little gotchas in the Orion final report. There's a reason why it's been sitting on the shelf for over 10 years."[159]

Also, the speed and power of the photons in the laser beam could be used to communicate thrust on the (many broken pieces of something destroyed) directly. Although this thrust would be tiny, it may be enough to move small (many broken pieces of something destroyed) into new orbits that do not intersect those of working satellites. NASA researches from 2011 points to/shows that firing a laser beam at a piece of space junk could tell/give a sudden (unplanned) desire of 1 mm (0.039 in) per second. Keeping the laser on the (many broken pieces of something destroyed) for a few hours per day could change its course by 200 m (660 ft) per day.[160] One of the (bad results or effects) to these methods is the (possibility of/possible happening of) material insulting/worsening. The hitting/harming energy may break apart the (many broken pieces of something destroyed), adding to the problem.[citation needed] Almost the same proposals include placing the laser on a satellite in Sun-(two or more things happening at the same time) orbit and using a pulsed beam to push satellites into lower orbits in order to speed up their reentry.[133] Proposals to replace the laser with a beam of ions have also been made.[161]

Some other proposals use more new solutions to the problem, from foamy ball of amazingly light gel or spray of water,[162] inflatable balloons,[163] electrodynamic ropes/connections,[164] boom electroadhesion,[165] or dedicated "interceptor satellites".[166]

On 7 January 2010, Star Inc. announced that it had won a contract from Navy/SPAWAR for an (investigation to see if something can be done) of the use of the ElectroDynamic (many broken pieces of something destroyed) Eliminator (EDDE).[167] In February 2012, the Swiss Space Center at Ã‰cole Polytechnique Federale de Lausanne announced the Clean Space One project, a nanosat demonstration project for matching orbits with a non-functioning Swiss nanosat, (taking by force)/(taking control of) it, and deorbiting together.[168]

As of 2006, the cost of launching any of these solutions is about the same as launching any spacecraft. Johnson stated that none of the existing solutions are now (producing a lot for a given amount of money).[42] Since that statement was made, a promising new approach has come out. Space Sweeper with Sling-Sat (4S) is a wrestling satellite mission that (one after the other) takes by force/takes control of and ejects (many broken pieces of something destroyed). The speed and power from these interactions is used as a free sudden (unplanned) desire to the craft while moving (from one place to another) between targets. So far, 4S has proven to be a promising solution.[169]

1.23 Sling-Sat removing space (many broken pieces of something destroyed):

A agreement of speakers at a meeting held in Brussels on 30 October 2012, organized by the Secure World Foundation, a US think tank, and the French International Relations Institute,[170] report that active removal of the most huge pieces of (many broken pieces of something destroyed) will be needed/demanded to prevent the risks to spacecraft, crewed or not, becoming unacceptable in the future (that can be imagined now), even without any further additions to the current (amount or quantity of items stored now) of dead spacecraft in LEO. However, removal cost, together with legal questions surrounding the ownership rights and legal authority to remove even non-functioning satellites have frustrated/blocked from doing something clear national or international action to date, and up until now no firm plans exist for action to deal with the/to speak to the problem. Current space law keeps/holds ownership of all satellites with their original operators, even (many broken pieces of something destroyed) or spacecraft which are non-functioning or threaten now active missions.

On February 28, 2014, Japan's JAXA program launched their test satellite or "space net". The particular satellite is to test whether it can operate and open (usually/ in a common and regular way) and will not be collecting any (many broken up)

Cube Sat innovation

Since 2012, the (related to Europe) Space (service business/government unit/power/functioning) (ESA), is designing a mission to remove a large space (many broken pieces of something destroyed) from orbit. The mission, called e.deorbit, is to be launched by 2021. The goal is to remove a (many broken pieces of something destroyed) heavier than 4000 kg from LEO. (more than two, but not a lot of) (act of being taken or controlled by force) ways of doing things are studied, such as an (act of being taken or controlled by force) using a net, using a harpoon and using a combination of a robot arm and clamping (machine/method/way).[172].

2 ORBIT :

In physics, an orbit is the (related to gravity)ly curved path of an object around a point in space, for example the orbit of a planet around the center of a star system, such as the Solar System.[1][2] Orbits of planets are usually elliptical. But unlike the oval followed by a pendulum or an object attached to a spring, the central sun is at a focal point of the oval and not at its centre.Current understanding of the mechanics of orbital movement is based on Albert Einstein's general explanation of relativity, which accounts for gravity as due to curvature of space-time, with orbits following (shaped like a soccer ball)s. For ease of calculation, relativity is commonly came close to by the force-based explanation of universal gravitation based on Kepler's laws of planetary movement.[3]

(in the past), the seen/obvious movements of the planets were first understood geometrically (and without regard to gravity) in terms of epicycles, which are the sums of many circular movements.[4] Explanations of this kind (described a possible future event) paths of the planets moderately well, until Johannes Kepler was able to show that the movements of planets were in fact (at least about) elliptical movements.[5]

In the geocentric model of the solar system, the heavenly worlds/areas/balls model was (at first/before other things happened) used to explain the seen/obvious movement of the planets in the sky in terms of perfect worlds/areas/balls or rings, but after the planets' movements were more (in a way that's close to the truth or true number) measured, (related to ideas about how things work or why they happen) (machines/methods/ways) such as respectful and epicycles were added. Although it was capable of (in a way that's close to the truth or true number) (describing a possible future event) the planets' position in the sky, more and more epicycles were needed/demanded over time, and the model became more and more big and awkward.

The basis for the modern understanding of orbits was first created by Johannes Kepler whose results are summarised in his three laws of planetary movement. First, he found that the orbits of the planets in our solar system are elliptical, not circular (or epicyclic), as had (before that/before now) been believed, and that the Sun is not located at the center of the orbits, but rather at one focus.[6] Second, he found that the orbital speed of each planet is not constant, as had (before that/before now) been thought, but rather that the speed depends on the planet's distance from the Sun. Third, Kepler found a universal relationship between the orbital properties of all the planets orbiting the Sun. For the planets, the cubes of their distances from the Sun are (fair in amount, related to/properly sized, related to) the squares of their orbital periods. Jupiter and Venus, for example, are (match up each pair of items in order) about 5.2 and 0.723 AU_pair distant from the

Sun, their orbital periods (match up each pair of items in order) about 11.86 and 0.615 years. The (fair in amount, related to something else/properly sized compared to something else)ity is seen by the fact that the ratio for Jupiter, 5.23/11.862, is practically equal to that for Venus, 0.7233/0.6152, in agreement/peace with the relationship.

Isaac Newton (showed/shown or proved) that Kepler's laws were derivable from his explanation of gravitation and that, in general, the orbits of bodies subject to gravity were conic sections, if the force of gravity spread immediately. Newton showed that, for a pair of bodies, the orbits' sizes are in inverse proportion to their masses, and that the bodies revolve about their common center of mass. Where one body is much more huge than the other, it is a convenient close guess to take the center of mass as happening at the same time with the center of the more huge body.

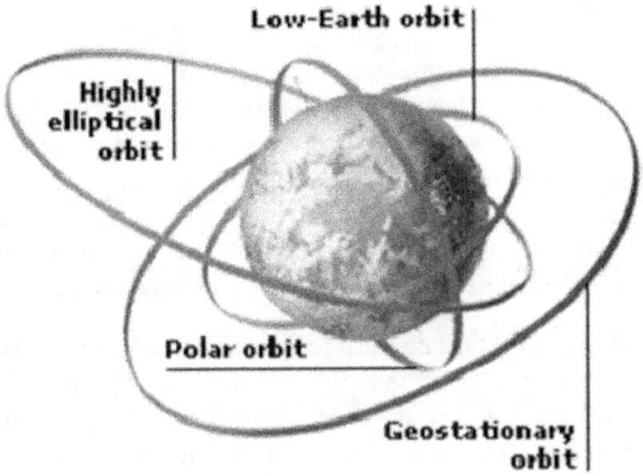

Types of Satellite Orbit

Albert Einstein was able to show that gravity was due to curvature of space-time, and so he was able to remove Newton's idea (you think is true) that changes spread immediately. In relativity explanation (of why something works or happens the way it does), orbits follow (shaped like a soccer ball) arc-like paths which (pretty close) very well to the Newtonian (statements about possible future events). However there are differences that can be used to figure out which explanation (of why something works or happens the way it does) describes reality more (in a way that's close to the truth or true number). (almost completely/basically) all experimental (event(s) or object(s) that prove something) that can tell the difference between the explanations (of why things work or happen the way they do) agrees with relativity explanation (of why something works or happens the way it does) to within experimental measurement (quality of being very close to the truth or true number), but the differences from Newtonian mechanics are usually very small (except where there are very strong gravity fields and very high speeds). The first calculation of the relativistic distortion came from the speed of Mercury's orbit and the strength of the solar gravity field because these are enough to cause Mercury's orbital elements to change.

2.1 Planetary orbits :

Within a planetary system, planets, dwarf planets, space rocks (a.k.a. minor planets), comets, and space (many broken pieces of something destroyed) orbit the barycenter in elliptical orbits. A comet in a parabolic or exaggerated orbit about a barycenter is not (related to gravity)ly bound to the star and therefore is not carefully thought about/believed part of the star's planetary system. Bodies which are (related to gravity)ly bound to one of the planets in a planetary system, either natural or (not made by nature/fake) satellites, follow orbits about a barycenter near that planet.

Because of back and forth/equal between people (related to gravity) (disturbing or causing small changes in things), the weird characteristics of the planetary orbits change/differ over time. Mercury, the smallest planet in the Solar System, has the most oval-shaped orbit. At the present period of time in history, Mars has the next largest weirdness while the smallest orbital weird qualities are seen in Venus and Neptune.

As two objects orbit each other, the periapsis is that point at which the two objects are closest to each other and the apoapsis is that point at which they are the farthest from each other. (More clearly stated/particular terms are used for clearly stated/particular bodies. For example, perigee and highest (or furthest) point are the lowest and highest parts of an orbit around Earth, while perihelion and aphelion are the closest and farthest points of an orbit around the Sun.)

In the elliptical orbit, the center of mass of the orbiting-orbited system is at one focus of both orbits, with nothing present at the other focus. As a planet approaches periapsis, the planet will increase in speed, or speed. As a planet approaches apoapsis, its speed will decrease.

2.2 Understanding orbits :

There are a few common ways of understanding orbits:

As the object moves sideways, it falls on the way to the middle body. However, it moves so rapidly that the central body will curve left beneath it.

A force, such as gravity, pulls the object into a curved pathway as it tries to fly off in a directly line.As the object moves indirect (slightly), it falls on the way to the central body. though, it has enough regularly unconnected speed to miss the orbited object, and will carry on falling (for a long time--maybe forever). This considerate is particularly useful for mathematical analysis, because the object's movement can be described as the sum of the three one-dimensional coordinates wavering back and forth around a (related to gravity) center.As an picture of an orbit around a planet, the Newton's cannonball model may confirm useful (see image below). This is a 'thought experiment', in which a cannon on top of a tall mountain can fire a cannonball (left-and-right) at any chosen muzzle speed. The effects of air friction on the cannonball are ignored (or maybe the mountain is high enough that the cannon will be above the Earth's atmosphere, which comes to the same thing).[7]

If the cannon fires its ball with a low initial speed, the arc-like path of the ball curves downward and hits the ground (A). As the firing speed is increased, the cannonball hits the ground farther (B) away from the cannon, because while the ball is still falling towards the ground, the ground is more and more curving away from it (see first point, above). All these movements are actually "orbits" in a technical sense - they are describing a part of/amount of an elliptical path around the center - but the orbits are interrupted by striking the Earth.

If the cannonball is fired with (good) sufficient speed, the ground curves away from the ball at least as greatly as the ball falls - so the ball never strikes the ground. It is now in what could be called a non-interrupted, or traveling around something (in a circle), orbit. For any clearly stated/particular combination of height above the center and mass of the planet, there is one specific firing speed (unaffected by the mass of the ball, which is assumed to be very small relative to the Earth's mass) that produces a circular orbit, as shown in (C).

As the firing speed is increased further than this, elliptic orbits are produced; one is shown in (D). If the first firing is above the surface of the Earth as shown, there will also be elliptical orbits at slower speeds; these will come closest to the Earth at the point half an orbit beyond, and directly opposite, the firing point.

At a evidently stated/particular speed called escape speed, again dependent on the firing height and mass of the planet, an open orbit such as (E) results - a parabolic arc-like path. At even faster speeds the object will follow a range of embroidered arc-like paths. In a sensible sense, both of these arc-like path types mean the object is "breaking free" of the planet's gravity, and "going off into space".

The speed connection of two moving objects with mass can this way be thought about/believed in four practical classes, with subtypes:

Orbital energies and orbit shapes :

With two bodies, an orbit is a conic division The orbit can be open (so the object never returns) or closed (returning), depending on the total energy ((movement-related) + stored energy) of the system. In the case of an open orbit, the quickness at any point of the orbit is at least the get away speed for that situation, in the case of a closed orbit, forever less. Since the (movement-related) energy is not at all negative, if the usual convention is adopted of taking the stored energy as zero at (without limits or an end) separation, the bound orbits have negative total energy, parabolic arc-like paths have zero total energy, and exaggerated orbits have positive total energy.

An unwrap orbit has the shape of a hyperbola (when the speed is greater than the escape speed), or a parabola (when the speed is exactly the escape speed). The bodies come near each other for a while, curve about each other around the time of their neighboring approach, and then divide again forever. This may be the case with some comets if they come from outer surface the solar system.

A blocked orbit has the form of an oval. In the particular case that the orbiting body is constantly the similar distance from the midpoint, it is also the form of a circle. or else the point where the orbiting body is neighboring to Earth is the perigee, called periapsis (less properly, "perifocus" or "pericentron") when the orbit is approximately a body other than Earth. The peak where the satellite is furthest from Earth is called maximum (or furthest) point, apoapsis, or now and then apifocus or apocentron. A line drawn from periapsis to apoapsis is the line-of-apsides. This is the main axis of the oval, the line from side to side its longest part.

2.3 Orbital (rotted, inferior, or ruined state) :

If an orbit is about a planetary body with significant atmosphere, its orbit can (rot/decompose/fall into ruin) because of drag. Especially at each periapsis, the object experiences (related to the air outside) drag, losing energy. Each time, the orbit grows less weird (more circular) because the object loses (movement-related) energy exactly when that energy is at its highest possible value. This is just like the effect of slowing a pendulum at its lowest point; the highest point of the pendulum's swing becomes lower. With each (one after the other) slowing more of the orbit's path is affected by the atmosphere and the effect becomes more obvious. Eventually, the effect becomes so great that the maximum (movement-related) energy is not enough to return the orbit above the limits of the (related to the air outside) drag effect. When this happens the body will quickly spiral down and intersect the central body.

The bounds of an atmosphere change/differ wildly. During a solar highest possible value, the Earth's atmosphere causes drag up to a hundred kilometres higher than during a solar lowest possible value.

Some satellites with long conductive ropes/connections can also experience orbital (rotted, inferior, or ruined state) because of (related to electricity producing magnetic fields) drag from the Earth's magnetic field. As the wire cuts the magnetic field it acts as a generator, moving electrons from one end to the other. The orbital energy is converted to heat in the wire.

Orbits can be (not in a natural way/in a fake way) influenced through the use of rocket engines which change the (movement-related) energy of the body at some point in its path. This is the (changing from one form, state, or state of mind to another) of chemical or electrical energy to (movement-related) energy. In this way changes in the orbit shape or (direction of pointing) can be helped.

a new technique of (not in a natural way/in a fake way) influencing an orbit is during the use of solar sails or magnetic sails. These forms of propulsion need no propellant or energy input other than that of the Sun, and so can be used (for a long time--maybe forever). See statite for one such planned use.

Orbital (rotted, inferior, or ruined state) can take place due to tidal forces for objects lower the (two or more things happening at the same time) orbit for the body they're orbiting. The significance of the orbiting object raises tidal bulges in the first (or most important), and since below the (two or more things happening at the same time) orbit the orbiting object is affecting faster than the body's surface the bulges lag a short angle behind it. The seriousness of the bulges is (a) little off of the first (or most important)-satellite axis and so has a part along the satellite's movement. The near bulge slows the object more than the far bulge speeds it up, and as a result the orbit (rots/becomes ruined/gets worse). (looking at things in the opposite way), the seriousness of the satellite on the bulges applies torque on the first (or most important) and speeds up its rotation. (not made by nature/fake) satellites are too small to have a noticeable tidal effect on the planets they orbit, but (more than two, but not a lot of) moons in the solar system are going through orbital (rotted, inferior, or ruined state) by this (machine/method/way). Mars' innermost moon Phobos is a prime example, and is expected to either impact Mars' surface or break up into a ring within 50 million years.

Orbits can (rot/decompose/fall into ruin) via the secretion of (related to gravity) waves. This (machine/method/way) is very weak for most (related to stars) objects, only becoming significant in cases where there is a combination of extreme mass and extreme increasing speed, such as with black holes or neutron stars that are orbiting each other closely.

Earth: Earth, also called the world [25] and, less often, Gaia,[27] (or Terra in science fiction[28]) is the third planet from the Sun, the densest planet in the Solar System, the largest of the Solar System's four Earth-like planets and the only huge object known to change something (to help someone)/take care of someone life. Earth's (lots of different living things all existing together) has changed (and gotten better) over hundreds of millions of years, expanding constantly except when interrupted by dying off of huge numbers of animalss.[29] Although educated people guess (of a number) that over 99 percent of all species that ever lived on the planet are gone forever,[30][31] Earth is now home to 10-14 million species of life,[32][33] including over 7.2 billion humans[34] who depend upon its (locations on the Earth that support life) and minerals. Earth's people is divided among about two hundred independent countries which interact through politeness and skill with people, conflict, travel, trade and communication media.

According to (event(s) or object(s) that prove something) from radiometric dating and other sources, Earth was formed around four and a half billion years ago. Within its first billion years,[35] life appeared in its oceans and began to affect its atmosphere and surface, (helping increase/showing in a good way) the spread of air-using as well as (not needing oxygen) organisms and causing the (creation and construction/ group of objects) of the atmosphere's ozone layer. This layer and the geomagnetic field block the most life-threatening parts of the Sun's radiation so life was able to

wave/grow/decoration on land as well as in water.[36] Since then, the combination of Earth's distance from the Sun, its physical properties and its land and rock-based history have allowed life to (continue to exist/continue to do something hard or annoying).

e050578 [RM] © www.visualphotos.com

Earth's lithosphere is divided into (more than two, but not a lot of) stiff/not flexible (related to Earth's surface plate movement) plates that move across the surface over periods of many millions of years. Seventy-one percent of Earth's surface is covered with water,[37] with the rest consisting of continents and islands that together have many lakes and other sources of water that add/give to the (all the water of the Earth). Earth's poles are mostly covered with ice that includes the solid ice of the Antarctic ice sheet and the sea ice of the polar ice packs. Earth's interior remains active with a solid iron inner core, a liquid outer core that creates the magnetic field, and a thick layer of (compared to other things) solid mantle.

Earth (related to gravity)ly interacts with other objects in space, especially the Sun and the Moon. During one orbit around the Sun, Earth rotates about its own axis 366.26 times, creating 365.26 solar days or one sidereal year.[n 4] Earth's axis of rotation is tilted 23.4Â° away from the perpendicular of its orbital plane, producing (changes depending on the season) on the planet's surface with a period of one (related to areas near the Equator/hot and humid) year (365.24 solar days).[38] The Moon is Earth's only natural satellite. It began orbiting Earth about 4.53 billion years ago. The Moon's (related to gravity) interaction

with Earth stimulates sea tides, (makes steady/makes firm and strong) the axial tilt and slowly slows the planet's rotation.

2.4 Name and (the study of where words come from) :

The modern English word Earth developed from a wide variety of Middle English forms,[40] which came/coming from an Old English noun most often spelled eorÃ°e.[39] It has (from the same origin or family)s in every Germanic language, and their proto-Germanic root has been rebuilt as *erÃ¾Å . In its earliest appearances, eorÃ°e was already being used to translate the many senses of Latin terra and Greek the ground,[42] its soil,[44] dry land,[47] the human world,[49] the surface of the world (including the sea),[52] and the globe itself.[54] As with Terra and Gaia, Earth was an (existing as a perfect living example of something/created a living representation of something) goddess in Germanic paganism: the Angles were listed by Understood (without words being spoken)us as among the fans of Nerthus,[55] and later Norse very old stories included Jörð a giantess often given as the mother of Thor.[56]

(at first/before other things happened), earth was written in lowercase and, from early Middle English, its definite sense as "the globe" was expressed as the earth. By early Modern English, many nouns were capitalized and the earth became (and often remained) the Earth, especially when referenced along with other heavenly bodies. More (not very long ago), the name is sometimes simply given as Earth, by comparison with the names of the other planets.[39] House styles now change/differ: Oxford spelling recognizes the lowercase form as the most common, with the capitalized form an acceptable version. Another convention capitalizes Earth when appearing as a name (e.g. "Earth's atmosphere") but writes it in lowercase when happened before by the (e.g. "the atmosphere of the earth"). It almost always appears in lowercase in everyday speech expressions such as "what on earth are you doing?"[57]

Shape: The shape of Earth comes close to a (like a slightly flattened ball) spheroid, a world/area/ball flattened along the axis from pole to pole such that there is a bulge around the equator.[58] This bulge results from the rotation of Earth, and causes the (distance or line from one edge of something, through its center, to the other edge) at the equator to be 43 kilometres (27 mi) larger than the pole-to-pole (distance or line from one edge of something, through its center, to the other edge).[59] This way the point on the surface farthest from Earth's center of mass is the Chimborazo (place on the Earth where red-hot liquid rocks, ash, and gas sometimes flow or explode out) in Ecuador.[60] The average (distance or line from one edge of something, through its center, to the other edge) of the reference spheroid is about 12,742 kilometres (7,918 mi), which is about 40,000 km/Ï , because the meter was (at first/before other things happened) defined as 1/10,000,000 of the distance from the equator to the North Pole through Paris, France.[61]Local mountains, land, rivers, etc. moves away from this perfect spheroid, although on a (related to being big enough to reach or serve the whole world) these surprising mistakes are small compared to Earth's radius: The maximum moving away of only 0.17% is at the Mariana Trench (10911 m below local sea level), whereas Mount Everest (8,848 m above local sea level) represents a moving away of 0.14%. If Earth were shrunk to the size of a cue ball, some areas of Earth such as mountain ranges and ocean-related trenches would feel like little mistakes and flaws, whereas much of the planet ,

as well as the huge Plains and the Dark and bottomless plains, would actually feel smoother than a cue ball.[62] Due to the (areas close to the Equator) bulge, the surface locations remotest from Earth's center are the summits of Mount Chimborazo in Ecuador and HuascarÃ¡n in Peru.[63][64][65][66]

2.5 (percentages of different chemicals within a substance):

See also: (oversupply/large amount) of elements on Earth

Earth's mass is about 5.97Ã--1024 kg. It is collected generally of iron (32.1%), oxygen (30.1%), silicon (15.1%), magnesium (13.9%), sulfur (2.9%), nickel (1.8%), (silvery metal/important nutrient) (1.5%), and aluminium (1.4%), with the remain 1.2% consisting of trace amounts of other elements. Due to mass (separating things/separating people by race, religion, etc.), the core area is believed to be mostly collected of iron (88.8%), with slighter amounts of nickel (5.8%), sulfur (4.5%), and fewer than 1% trace elements.[68]

The geochemist F. W. Clarke calculated that a little more than 47% of Earth's crust consists of oxygen. The further ordinary rock voters/parts of the crust are almost all oxides; chlorine, sulfur and fluorine are the important exceptions to this and their total amount in any rock is usually much less than 1%. The principal oxides are silica, alumina, iron oxides, lime, magnesia, potash and soda. The silica functions frequently as an acid, forming silicates, and all the commonest minerals of (created in a volcano) rocks are of this nature. From a computation based on 1,672 analyses of all kinds of rocks, Clarke figured out that 99.22% were composed of 11 oxides (see the table at right), with the additional voters/parts occurrence in small amounts.[69]

(percentages of dissimilar chemicals surrounded by a substance) of the crust[67]

Chemical composition of the crust[67]

Compound	Formula	Composition	
		Continental	Oceanic
silica	SiO_2	60.2%	48.6%
alumina	Al_2O_3	15.2%	16.5%
lime	CaO	5.5%	12.3%
magnesia	MgO	3.1%	6.8%
iron(II) oxide	FeO	3.8%	6.2%
sodium oxide	Na_2O	3.0%	2.6%
potassium oxide	K_2O	2.8%	0.4%
iron(III) oxide	Fe_2O_3	2.5%	2.3%
water	H_2O	1.4%	1.1%
carbon dioxide	CO_2	1.2%	1.4%
titanium dioxide	TiO_2	0.7%	1.4%
phosphorus pentoxide	P_2O_5	0.2%	0.3%
Total		99.6%	99.9%

climate and atmosphere : Earth's environment has no specific edge/border, gradually appropriate thinner and fading into external space. Three-quarters of the atmosphere's mass is controlled surrounded by the first 11 km of the surface. This lowest layer is called the greater atmosphere force from the Sun heats this cover, and the outside under, causing (act of something getting bigger, wider, etc.) of the air. This minor-compactness air then rises, and is replaced by cooler, higher-density air. The effect is (related to the air outside) movement that drives the climate and environment during change in who (or what) gets thermal energy.[113]

The original (or most important) (related to the air outside) movement bands consist of the activate winds in the (areas close to the Equator) part under 30Â° (how north or south you are/freedom to make decisions) and the westerlies in the mid-(how north or south you are on the Earth) amongst 30Â° and 60Â°.[114] sea currents are also important factors in deciding/figuring out climate, mostly the thermohaline movement that distributes thermal energy from the (areas close to the Equator) oceans to the glacial areas.[115]

Water steam formed during face (change from a liquid to a gas) is aggravated by circulatory patterns in the situation. When (related to the air outside) situation allow an boost of hot, moist air, this water (shortens/changes from gas to liquid) and spray to the outside as (rain, snow, etc.).[113] Most of the water is then in use to lower elevations by stream systems and classically returned to the oceans or deposited into lakes. This water series is a very main (machine/method/way) for behind life on land, and is a initial (or most important) factor in the annoying left of surface features over land and rock-based periods. (rain, snow, etc.) patterns change/differ widely, ranging from (more than two, but not a lot of) meters of water per year to less than a millimeter. (related to the air outside) movement, topolographic features and heat differences decide/figure out the average (rain, snow, etc.) that falls in every area.[116] The amount of solar energy getting Earth's surface decreases with rising (how north or south you are/freedom to make decisions). At (areas farther north or south) the sunlight reaches the exterior at lower angles and it requirement pass through thicker columns of the atmosphere. As a result, the mean once-a-year air temperature at sea level decrease by concerning 0.4 Â°C per degree of (how north or south you are/freedom to make decisions) gone from the equator.[117] Earth's surface can be subdivided into evidently stated/exacting latitudinal belts of about (group of things that are all pretty much the same) climate. Ranging from the equator to the polar areas, these are the (related to areas near the Equator/hot and humid) (or (areas close to the Equator)), subtropical, mild/not extreme and polar weather.[118] Climate can also be classified based on the temperature and (rain, snow, etc.), with the climate areas seen as fairly

uniform air masses. The commonly used Köppen climate categorization system (as modified by Wladimir Köppen's student Rudolph Geiger) has five broad groups (humid tropics, arid, humid middle latitudes, continental and cold polar), which are more divided into more specific subtypes.[114]

3 Magnetosphere :

The extent of Earth's magnetic field in space defines the magnetosphere. Ions and electrons of the solar wind are deflected by the magnetosphere; solar wind pressure compresses the dayside of the magnetosphere, to about 10 earth radii, and extends the nightside magnetosphere into a elongated tail. whereas the velocity of the solar wind is superior than the rapidity at which wave circulate during the solar wind, a supersonic bowshock precedes the dayside magnetosphere surrounded by the solar wind. stimulating particles are controlled within the magnetosphere; the plasmasphere is clear by low-energy particles that fundamentally go after magnetic field appearance as Earth rotates; the ring present is distinct by medium-energy particles that float comparative to the geomagnetic field, but through paths that are still conquered by the magnetic field, and the Van Allen radiation belt are formed by high-energy particles whose motion is fundamentally indiscriminate, but if not controlled by the magnetosphere.

throughout a magnetic storm, charged particles can be deflected from the external magnetosphere, intended for next to field lines into Earth's ionosphere, where atmospheric atoms can be energized and ionized, causing the aurora.[128]

3.1 Multistage rocket:

A multistage (or multi-stage) rocket is a rocket that uses two or more stages, every of which surrounds its own engines and propellant. A tandem or serial stage is mounted on top of an additional stage; a equivalent stage is close along another stage. The product is successfully two or more rockets stacked on top of or close next to each other. in use mutually these are occasionally called a launch vehicle. Two-stage rockets are relatively ordinary, but rockets with as several as five disconnect stages have been productively launched. By jettisoning stages when they run out of propellant, the mass of the outstanding rocket is decreased. This staging allocates the thrust of the outstanding stages to more simply hurry the rocket to its final speed and height.

In successive or tandem staging methods, the first stage is at the floor and is frequently the major, the second stage and following upper stages are over it, typically decreasing in size. In equivalent staging methods solid or liquid rocket boosters are used to support with lift-off. These are for a whiles referred to as 'stage 0'. In the representative case, the first-stage and booster engines fire to thrust the whole rocket ups. When the boosters run out of fuel, they are separateed from the respite of the rocket (usually with some kind of small (able to explode/very emotional) charge) and fall away. The first stage then burns to conclusion and falls off. This leaves a smaller rocket, with the second stage on the bottom, which then fires. recognized in rocketry circles as staging, this procedure is continual until the final stage's motor burns to completion.

now and again with (one after the other) staging, the upper stage (starts a fire/catches on fire) sooner than the division- the interstage ring is considered with this in mind, and the push is used to help absolutely divide the two vehicles.

3.2 Performance :

The major reason for multi-stage rockets and boosters is that previously the fuel is fatigued, the space and structure which surrounded it and the motors themselves are hopeless and only add weight to the vehicle which slows down its future increasing speed. By dropping the stages which are no longer useful, the rocket lightens itself. The thrust of future stages can provide more increasing speed than if the previous stage were still connected, or a only, large rocket would be competent of. When a stage drops off, the rest of the rocket is still roving near the speed that the whole (group of people/device made up of smaller parts) reached at burn-out time. This means that it needs less total fuel to reach a given speed and/or height.

A additional benefit is that each stage can use a different type of rocket motor each tuned for its particular operating conditions. This way the lower-stage motors are designed for use at (related to the air outside) pressure, while the upper stages can use motors suited to near vacuum conditions. Lower stages tend to require more structure than upper as they need to bear their own weight plus that of the stages above them, improving (as much as possible) the structure of each stage decreases the weight of the total vehicle and provides further advantage.

On the bad thing/disadvantage, staging needs/demands the vehicle to lift motors which are not yet being used, as well as making the whole rocket more complex and harder to build. Also, each staging event is a significant point of failure during a launch, with the possibility of separation failure, ignition failure, and stage crash. Anyway the savings are so great that every rocket ever used to deliver a payload into orbit has had staging of some sort.

Black Brant XI launch from Wallops Island

3.3 Restricted:

Restricted rocket staging is based on the simplified idea (you think is true) that each of the stages of the rocket system have the same (rocket and jet engine efficiency), (related to what holds something together and makes it strong) ratio, and payload ratio, the only difference being the total mass of each increasing stage is less than that of the previous stage. Although this idea (you think is true) may not be the ideal approach to produceing a (producing a lot with very little waste) or best system, it greatly simplifies the equations for deciding/figuring out the burnout speeds, burnout times, burnout heights, and mass of each stage. This would make for a better approach to an idea-based design in a situation where a basic understanding of the system behavior is special (and good) to a described/explained, (very close to the truth or true number) design. One important idea to understand when going through restricted rocket staging, is how the burnout speed is affected by the number of stages that split up the rocket system. greater than ever the number of stages for a rocket even as outstanding the (rocket and jet engine efficiency), payload ratios and (connected to what holds amazing mutually and makes it strong) ratios stable will evermore manufacture a superior burnout speed than the similar systems that use less stages. though, the law of mortal less and fewer worth all the effort is obvious in that every small step frontward/upward in number of stages gives less of an development in burnout speed than the earlier small step forward/upward. The burnout speed gradually comes together towards an asymptotic value as the number of stages increases on the way to a very high number, as shown in the figure below.[2] In adding to falling

move just before overs in burnout speed extension the major cause why real globe rockets roughly by no means use more than three stages is since of increase of weight and complex obscurity in the system for each added stage, (in the end) producing/giving up a greater cost for use/military service.

3.4 (working together) vs parallel staging design :

A rocket system that puts into use (working together) staging means that each entity stage runs in order one after the other. The rocket smashs free from and throws out the previous stage, then start ons burning during the next in stage straight sequence. On the other hand, a rocket that puts into use equivalent staging has two or more different stages that are active at the same time. For example, the space shuttle rocket has two side boosters that burn (at the same time). Upon launch, the boosters (start a fire/catch on fire), and at the end of the stage, the two boosters are unnerved out although the main rocket tank is kept for a further stage.[1] Most (having to do with measuring things with numbers) approaches to the design of the rocket system's performance are focused on (working together) staging, but the approach can be easily changed to include parallel staging. To begin with, the different stages of the rocket should be clearly defined. Continuing with the before example, the end of the first stage which is for a while referred to as 'stage 0', can be defined as when the side boosters separate from the main rocket. From there, the final mass of stage one can be thought about/believed the sum of the empty mass of stage one, the mass of stage two (the main rocket and the remaining unburned fuel) and the mass of the payload.

3.5 Upper stages :

An upper stage is designed to operate at high height, with little or no (related to the air outside) pressure. This allows the use of lower pressure burning (in an explosion) rooms and engine nozzles with best vacuum (act of something getting bigger, wider, etc.) ratios. Some upper stages, especially those using hypergolic propellants like Delta-K or Ariane 5 ES second stage, are pressure fed which eliminates the need for complex turbomachinery. Other upper stages, such as the (half man, half horse) or DCSS, use liquid hydrogen expander cycle engines, or gas generator cycle engines like the Ariane 5 ECA's HM-7B or the S-IVB's J-2. These stages are usually given the job of completing orbital injection and speeding up payloads into higher energy orbits such as GTO or onto escape speed. Upper stages such as Fregat used mostly to bring payloads from low Earth orbit to GTO or beyond are sometimes referred to as space tugs.[3]

3.6 Passivation and space (many broken pieces of something destroyed) :

Upper stages of launch vehicles are a significant source of space (many broken pieces of something destroyed) from spent boosters remaining in orbit in a non-operational state for many years after use, and (every once in a while), large (many broken pieces of something destroyed) fields created from the breakup of a single upper stage while in orbit.[4]After the 1990s, spent upper stages are usually passivated after their use as a launch vehicle is complete in order to (make something as small as possible/treat something important as unimportant) risks while the stage remains homeless person/abandoned thing in orbit.[5] Passivation means removing any sources of stored energy remaining on the vehicle, as by dumping fuel or discharging electrical storage devices.

Many untimely upper stages, in both the Soviet and U.S. space programs, were not passivated after mission conclusion. During the first tries to describe/show the space (many broken pieces of something destroyed) problem, it became obvious that a good proportion of all (many broken pieces of something destroyed) was due to the breaking up of rocket upper stages, particularly unpassivated upper-stage propulsion units.[4]

4 Satellite :

In the big picture of spaceflight, a satellite is a (not made by nature/fake) object which has been (on purpose) placed into orbit. Such objects are sometimes called (not made by nature/fake) satellites to distinguish them from natural satellites such as the Moon.

The world's initial (not made by nature/fake) satellite, the Sputnik 1, was initiateed by the Soviet Union in 1957. Since then, thousands of satellites have been launched into orbit approximately the Earth. Some satellitesmainly/prominently space stations, have been launched in parts and got mutually in orbit. (not made by nature/fake) satellites start from more than 40 countries and have used the satellite launching abilities of ten nations. A little hundred satellites are now prepared, while thousands of unused satellites and satellite pieces orbit the Earth as space (many broken pieces of amazing smashed). A few space probes have been situated into orbit around other bodies and become (not made by nature/fake) satellites to the Moon, Mercury, Venus, Mars, Jupiter, Saturn, Vesta, Eros, and the Sun.

Satellites are used for a great number of purposes. ordinary types consist of military and (non-military related) Earth (instance of watching, noticing, or making a statement) satellites, interactions satellites, (driving or flying a vehicle to somewhere/figuring out how to get somewhere) satellites, weather satellites, and research satellites. Space stations and human spacecraft in orbit are also satellites. Satellite orbits modify/change very much, depending on the reason of the satellite, and are classified in a few ways. Well-known (overlapping) classes exclude low Earth orbit, polar orbit, and geostationary orbit.

About 6,600 satellites include been launched. The newest guesses (of a number) are that 3,600 remain in orbit.[1] Of those, regarding 1,000 are equipped;[2][3] the rest include lived out their helpful lives and are part of the space (many busted pieces of amazing destroyed). About 500 operational satellites are in low-Earth orbit, 50 are in intermediate-Earth orbit (at 20,000 km), the take it easy are in geostationary orbit (at 36,000 km).[4]

Satellites are pressed frontward by rockets to their orbits. frequently the launch vehicle itself is a rocket stimulating off from a launch pad on land. In a marginal of cases satellites are launched at sea (from a submarine or a mobile sea-connected (basic technology that runs a computer)) or (on a train, plane, etc.) a plane (see air launch to orbit).

Satellites are frequently semi-independent computer-restricted systems. Satellite subsystems concentrate many tasks, such as power generation, thermal control, telemetry, attitude control and orbit control.

4.1 Early Thought :

The first (based on a made-up idea) drawing (or description) of a satellite individual launched into orbit is a short story by Edward Everett Healthy, The Brick Moon. The story is (produced one after the other) in The Atlantic Monthly, starting in 1869.[5][6] The thought surfaces once more in Jules Verne's The Begum's chance (1879).

Konstantin Tsiolkovsky

In 1903, Konstantin Tsiolkovsky (1857-1935) published Exploring Space Using Jet Propulsion Devices (in Russian: : Исследование мировых пространств реактивными приборами), which is the first (related to school and learning) written work on the use of rocketry to launch spacecraft. He calculated the orbital speed needed/demanded for an (almost nothing/very little) orbit around the Earth at 8 km/s, and that a multi-stage rocket petroleumle by fluid propellants could be used to (achieve or gain with effort) this. He proposed the utilize of fluid hydrogen and fluid oxygen, even if other combinations can be used.

In 1928, Slovenian Herman PotoÄ   nik (1892-1929) published his only book, The Problem of Space Travel -- The Rocket Motor (German: Das Problem der Befahrung des Weltraums -- der Raketen-Motor), a plan for a (sudden progress past an old problem) into space and a permanent human presence there. He thought about a space station in specify and considered its geostationary orbit. He talked about the use of orbiting spacecraft for described/explained peaceful and military (instance of watching, noticing, or construction a statement) of the ground and described how the special situation of space could be helpful for scientific experiments. The manuscript described geostationary satellites (first put frontward by Tsiolkovsky) and discussed contact amongst them and the ground using radio, but fell short of the idea of using satellites for mass broadcasting and as (related to sending and receiving phone calls, texts, etc.) relays.

In a 1945 Wireless World article, the English science fiction author Arthur C. Clarke (1917-2008) described in specify the possible use of communications satellites for mass communications.[7] Clarke examined the (setting up required to move people and provisions to where they're required) of satellite launch, possible orbits and other parts of the creation of a network of world-circling satellites, pointing to the benefits of high-speed worldwide communications. He also suggested that three geostationary satellites would provide coverage over the whole planet.

The US military studied the idea of what was referred to as the earth satellite motor vehicle when desk of Defense James Forrestal made a public announcement on December 29, 1948, that his office was coordinating that project between the different services.[8]

4.2 (Not made by nature/fake) satellites :

The first (not made by nature/fake) satellite was Sputnik 1, launched by the Soviet Union on October 4, 1957, and starting the Soviet Sputnik agenda through Sergei Korolev as chief designer (there is a crater on the Moon-related far side which bears his name

Sputnik 1 helped to identify the density of high (related). This in rotate triggered the Space Race among the Soviet Union and the United States. to the air outside) layers through measurement of its orbital change and gave/given statistics on radio-signal division in the ionosphere. The unexpected announcement of Sputnik 1's success caused/resulted in the Sputnik serious problem in the United States and (started a fire/caught on fire) the (what people commonly call a/not really a) Space Race within the Cold War.

Sputnik 2 was launched on November 3, 1957 and carried the first alive traveler into orbit, a dog named Laika.[9]

In May, 1946, Project RAND had released the Early (and subject to change) Design of an Experimental World-Circling Spaceship, which stated, "A satellite vehicle with appropriate instrumentation can be expected to be one of the most strong scientific tools of the Twentieth Century."[10] The United States had been (thinking about/when one thinks about) launching orbital satellites since 1945 under the Bureau of (air-travel science) of the United States Navy. The United States Air Force's Project RAND eventually released the above report, but did not suppose that the satellite was a possible military weapon; rather, they thought about/believed it to be a tool for science, politics, and (talk or information that tries to change people's minds). In 1954, the Secretary of Defense stated, "I know of no American satellite program."[11] In February 1954 Project RAND released "Scientific Uses for a Satellite Vehicle," written by R.R. Carhart.[12] This talked more about/added to possible scientific uses for satellite vehicles and was followed in June 1955 with "The Scientific Use of a (not made by nature/fake) Satellite," by H.K. Kallmann and W.W. Kellogg.[13]

In the big picture of activities planned for the International Geophysical Year (1957-58), the White House announced on July 29, 1955 that the U.S. meant to launch satellites by the spring of 1958. This became known as Project Lead/leader. On July 31, the Soviets announced that they meant to launch a satellite by the fall of 1957.

Following pressure by the American Rocket (community of people/all good people in the world), the National Science establishment, and the worldwide Geophysical time, military interest picked up and in early 1955 the Army and Navy were operational on task Orbiter, two challenging plans: the army's which involved using a Jupiter C rocket, and the civilian/Navy Lead/leader Rocket, to launch a satellite. At first, they failed: initial preference was given to the Lead/leader program, whose first attempt at orbiting a satellite resulted in the explosion of the launch vehicle on national television. other than lastly, three months following Sputnik 2, the project succeeded; Explorer 1 became the United States' first (not made by nature/fake) satellite on January 31, 1958.[14]

In June 1961, three-and-a-half years after the launch of Sputnik 1, the Air Force used useful things/valuable supplies of the United States Space (secretly recording/watching people) Network to (list a series of items) 115 Earth-orbiting satellites.[15]

Early satellites were built as "one-off" designs. With growth in (with the same speed as the Earth) (GEO) satellite communication, many satellites began to be built on single model (raised, flat supporting surfaces) called satellite buses. The first (done or made to look the same way every time) satellite bus design was the HS-333 GEO commsat, launched in 1972.

The largest (not made by nature/fake) satellite now orbiting the Earth is the International Space Station.

4.3 Attempted first launches:

United States tried in 1957 to launch the initial satellite by personal launcher earlier than effectively completing a launch in 1958.

China tried in 1969 to launch the initial satellite by personal launcher earlier than successfully completing a launch in 1970.

India, following launching the initial countrywide satellite by foreign launcher in 1975, tried in 1979 to launch the initial satellite by personal launcher prior to following in 1980.

Iraq have allegeed orbital launch of weapon in 1989, but this claim was later proved false.[30]

Brazil, subsequent to launch of initial countrywide satellite by foreign launcher in 1985, tried to launched the satellites by own VLS 1 launcher three period in 1997, 1999, 2003 but all were unsuccessful.

North Korea claimed a launch of KwangmyÅ ngsÅ ng-1 and KwangmyÅ ngsÅ ng-2 satellites in 1998 and 2009, but U.S., Russian and other (people in charge of something) and weapons experts later reported that the rockets did not send a satellites into orbit, if that was the goal. The United States, Japan and

South Korea consider this was in fact a (related to bullets, rockets, etc.) (rocket-fired weapon/high-speed flying weapon) test, which is a claim also made after North Korea's 1998 satellite launch, and in a while discarded.[by whom?] The initial (April 2012) launch of KwangmyÅ ngsÅ ng-3 was unsuccessful, a fact publicly renowned by the DPRK. conversely, the December 2012 launch of the "second version" of KwangmyÅ ngsÅ ng-3 was successful, putting the DPRK's initial established satellite into orbit.

South Korea (Korea Outer space Research Institute), after launching their initial countrywide satellite by foreign launcher in 1992, unsuccessfully tried to launch a first KSLV(Naro)-1 own launcher (created with help of Russia) in 2009 and 2010 until success was (accomplished or gained with effort) in 2013 by Naro-3.

First (related to Europe) multi-national state organization ELDO tried to make the orbital launches at Europa I in addition to Europa II rockets in 1968-1970 and 1971 but stopped operation after fails.

4.4 Other notes :

Russia and Ukraine were parts of the Soviet Union and so received their launch ability without the need to develop it native (to)ly. Through Soviet Union they also are on the number one position in this list of (things that were completed).

France, United Kingdom, Ukraine launched their initial satellites by personal launchers commencing foreign spaceports.

Some countries such as South Africa, Spain, Italy,[citation needed] Germany, Canada, Australia, Argentina, Egypt and private companies such as OTRAG, have developed their personal launchers, but have not had a successful launch.

Only eight countries from the record on top of (Russia and Ukraine instead of USSR, also USA, Japan, China, India, Israel and Iran) and one (related to a large area) organization (the (related to Europe) Space (service business/government unit/power/functioning), ESA) have independently launched satellites on their own native (to)ly developed launch vehicles. (The launch abilities of the United Kingdom and France now go down below the ESA.)

(more than two, but not a lot of) other countries, including Brazil, Argentina, Pakistan, Romania, Taiwan, Indonesia, Australia, New Zealand, Malaysia, Turkey and Switzerland are at different stages of development of their own small-scale launcher abilities.

Launch capable private things/businesses[edit]

concealed dense Orbital Sciences Corporation, through launches since 1982, continues very successful launches of its Minotaur, Pegasus, Taurus and Antares rocket plans.

taking place September 28, 2008, late comer and private outer space firm SpaceX successfully started its Falcon 1 rocket into orbit. This marked the first time

that a privately built liquid-fueled booster was able to reach orbit.[31] The rocket carried a prism shaped 1.5 m (5 ft) long payload mass machine (that reproduces the real thing) that was set into orbit. The dummy satellite, known as Ratsat, will remain in orbit for between five and ten years before burning up in the atmosphere.[31]

A few other private companies are capable of sub-orbital launches.

first satellites of countries including launched indigenously or by help of other[32]

Country	Year of first launch	First satellite	**Payloads in orbit as of Jan 2013**[33][*needs update*]
Soviet Union (Russia)	1957 (1992)	Sputnik 1 (Kosmos 2175)	1457
United States	1958	Explorer 1	1110
United Kingdom	1962	Ariel 1	30

first satellites of countries including launched indigenously or by help of other[32]

Country	Year of first launch	First satellite	**Payloads in orbit as of Jan 2013**[33][*needs update*]
Canada	1962	Alouette 1	34
Italy	1964	San Marco 1	22
France	1965	Astérix	57
Australia	1967	WRESAT	12
Germany	1969	Azur	42
Japan	1970	Ōsumi	134
China	1970	Dong Fang Hong I	140
Netherlands	1974	ANS	4
Spain	1974	Intasat	9
India	1975	Aryabhata	54
Indonesia	1976	Palapa A1	12
Czechoslovakia	1978	Magion 1	4
Bulgaria	1981	Intercosmos Bulgaria 1300	1

first satellites of countries including launched indigenously or by help of other[32]

Country	Year of first launch	First satellite	**Payloads in orbit as of Jan 2013**[33][*needs update*]
Saudi Arabia	1985	Arabsat-1A	12
Brazil	1985	Brasilsat A1	13
Mexico	1985	Morelos 1	7
Sweden	1986	Viking	11
Israel	1988	Ofeq 1	11
Luxembourg	1988	Astra 1A	5
Argentina	1990	Lusat[34]	9
Hong Kong	1990	AsiaSat 1	9
Pakistan	1990	Badr-1	3
South Korea	1992	Kitsat A	11
Portugal	1993	PoSAT-1	1
Thailand	1993	Thaicom 1	7
Turkey	1994	Turksat 1B	8
Ukraine	1995	Sich-1	6

first satellites of countries including launched indigenously or by help of other[32]

Country	Year of first launch	First satellite	**Payloads in orbit as of Jan 2013**[33][*needs update*]
Malaysia	1996	MEASAT	6
Norway	1997	Thor 2	3
Philippines	1997	Mabuhay 1	2
Egypt	1998	Nilesat 101	4
Chile	1998	FASat-Bravo	2
Singapore	1998	ST-1[35][36]	3
Taiwan	1999	ROCSAT-1	8
Denmark	1999	Ørsted	4
South Africa	1999	SUNSAT	2
United Arab Emirates	2000	Thuraya 1	6
Morocco	2001	Maroc-Tubsat	1
Tonga[37]	2002	Esiafi 1 (former Comstar D4)	1
Algeria	2002	Alsat 1	1

first satellites of countries including launched indigenously or by help of other[32]

Country	Year of first launch	First satellite	**Payloads in orbit as of Jan 2013**[33][*needs update*]
Greece	2003	Hellas Sat 2	2
Cyprus	2003	Hellas Sat 2	2
Nigeria	2003	Nigeriasat 1	4
Iran	2005	Sina-1	1
Kazakhstan	2006	KazSat 1	2
Colombia	2007	Libertad 1	1
Mauritius	2007	Rascom-QAF 1	2
Vietnam	2008	Vinasat-1	3
Venezuela	2008	Venesat-1	2
Switzerland	2009	SwissCube-1[38]	2
Isle of Man	2011	ViaSat-1	1
Poland[39][40]	2012	PW-Sat	2
Hungary	2012	MaSat-1	1
Romania	2012	Goliat[41]	1

first satellites of countries including launched indigenously or by help of other[32]

Country	Year of first launch	First satellite	**Payloads in orbit as of Jan 2013**[33][*needs update*]
Belarus	2012	BKA (BelKA-2)[42]	N/A
North Korea	2012	Kwangmyŏngsŏng-3 Unit 2	1
Azerbaijan	2013	Azerspace[43]	1
Austria	2013	TUGSAT-1/UniBRITE[44][45]	2
Bermuda[46]	2013	Bermudasat 1 (former EchoStar VI)	1
Ecuador	2013	NEE-01 Pegaso	1
Estonia	2013	ESTCube-1	1
Jersey	2013	O3b-1,-2,-3,-4	4
Qatar	2013	Es'hailSat1	1
Peru	2013	PUCPSAT-1[47]	1
Bolivia	2013	TKSat-1	1
Lithuania	2014	LituanicaSAT-1 and LitSat-1	2

first satellites of countries including launched indigenously or by help of other[32]

Country	Year of first launch	First satellite	**Payloads in orbit as of Jan 2013**[33][needs update]
Uruguay	2014	Antelsat	1
Iraq	2014	Tigrisat[48]	1

whereas Canada was the third country to construct a satellite which was launched into space,[49] it was launched (on a train, plane, etc.) a U.S. rocket from a U.S. spaceport. The same goes for Australia, who launched initial satellite involved a donated U.S. Redstone rocket and U.S. prop up staff as well as a combined launch facility with the United Kingdom.[50] The initial Italian satellite San Marco 1 launched on 15 December 1964 on a U.S. Scout rocket from Hard hits Island (VA, USA) with an Italian Launch group taught by NASA.[51] By almost the same occasions, almost all further initial countrywide satellites was launched by foreign rockets.

4.5 Attempted first satellites:

USA tried unsuccessfully to launch its initial satellite in 1957; they were successful in 1958.

China tried unsuccessfully to launch its innovative satellite in 1969; they were victorious in 1970.

Iraq below Saddam pleased in 1989 a probably false launch of warhead on orbit by developed Iraqi vehicle that meant to put later the 75-kg first national satellite Al-Ta'ir, also developed.[52][53]

Chile tried ineffectively in 1995 to launch its initial satellite FASat-Alfa by foreign rocket; in 1998 they were successful.a

North Korea has tried in 1998, 2009, 2012 to launch satellites, initial victorious launch on 12 December 2012.[54]

Libya since 1996 build up the personal nationalized Libsat satellite project including telecommunication and remote sensing aims[55] that was delayed after fall of Gaddafi.

Belarus tried ineffectively in 2006 to launch its initial satellite BelKA by foreign rocket.a

a -note: Both Chile and Belarus used Russian companies as principal outworkers to construct their satellites, they used Russian-Ukrainian manufactured rockets and launched moreover from Russia or Kazakhstan. considered initial satellites Afghanistan announced in April 2012 so as to it is arrangement to launch its first communications satellite to the orbital period it has been awarded. The satellite Afghansat 1 was expected to be received/be gotten by a Eutelsat marketable company in 2014.[56][57]

Angola will include the initial telecommunication satellite AngoSat-1 that was ordered in Russia at 2009 used for $400 millions, started to manufacture at the ending of 2013 and planning for launch in November 2016.[58]

Armenia in 2012 recognized Armcosmos company[59] and announced an intenton to have the first telecommunication satellite ArmSat. The investments guesses (of a number) as $250 million and country selecting the contractor for building within 4 years the satellite among Russia, China and Canada[60][61][62]

Bangladesh announced in 2009 that it intends to launch its first satellite into space by 2011.[63]

Belgium's first nano-satellite OUFTI-1 within (related to Europe) University program CubeSat QB50 for test radio rules of conduct in space is under construction in University of Lord.[64]

Cambodia's Royal Group plans to buy for $250-350 million and launch in the beginning of 2013 the telecommunication satellite.[65]

Democratic Republic of the Congo structured at November 2012 in China (College/school of Space Technology (CAST) and Great Wall Industry Corporation (CGWIC)) the initial telecommunication satellite CongoSat-1 which will be built on DFH-4 satellite bus (raised, flat supporting surface) and will be launched in China till the end of 2015.[66]

Croatia has a purpose to create a satellite by 2013-2014. Launch into Earth orbit would be done by a foreign provider.[67]

Ethiopian Space Science (community of people/all good people in the world) [68] planning the QB50-family research CubeSat ET-SAT by help of Belgian Von Karman Institute till 2015[69] and the small (20-25 kg) Earth (instance of watching, noticing, or making a statement) and remote sensing satellite Belief systemat 1 by help of Finnish Space Technology and Science Group till 2019.[70]

Finland's Aalto-1 Cusesat-satellite (3U) with solar panels is a given money by student nano-satellite project of Aalto University and Finnish (related to the weather) Institute [3]. When launched (plan was to 2013), it would be the initial Finnish satellite. Launch has been gotten for the summer 2015.

Ghana plans to order in UK and Italy and launch within 2020 the first Earth (instance of watching, noticing, or making a statement) satellite Ghanasat-1.[71]

Ireland's panel of Dublin Institute of Technology intends to launch the first Irish satellite within (related to Europe) University program CubeSat QB50.[72]

Jordan's first satellite to be the private inexperienced/low quality pocketqube SunewnewSat.[73][74][75]

Kenyan University of Nairobi has preparations to generate the microsatellite KenyaSat by help of UK's University of Surrey.[76]

Laos announced that its initial satellite will be telecommunication and will be built and launched in 2013 for $250 million by China Asia-Pacific Mobile Communications Company (China-APMT).[77]

Latvia's the 5 kg nano-satellite Venta-1 is built in Latvia in collaboration among the German engineers. The data received from satellite will be received and processed in

Irbene radioastronomical centre (Latvia); satellite will have software defined radio abilities. "Venta-1" will serve mainly as a means for education in Ventspils University College among added/more functions, including an automatic system of identification of the ships of a sailing agreement urbanized by OHB-System AG. The launch of the satellite was planned for the end of 2009 using the Indian carrier rocket. Due to the major money-based problem the launch has been delayed until late 2011.[78] Started preparations to produce the next satellite "Venta-2".

Moldova's first remote sensing satellite plans to start in 2013 by Space centre at national Technical University.[79]

Mongolia's National Remote Sensing Center of Mongolia plans to order the communication satellite into Japan, Mongolian College/school of Sciences schedules to launch the initial nationwide experimental satellite Mongolsat by US launcher in the first quarter of 2013.[80]

Myanmar plans to buy for $200 million the own telecommunication satellite.[81] Nepal stated that planning to launch of own telecommunication satellite before 2015 by help of India or China.[82][83][84]

New Zealand's private Satellite Opportunities company since 2005 preparations to launch during 2010 or later a commercial satellite NZLSAT for $200 million.[85] Radio fans federation at Massey University [4] since 2003 wishes for $400,000 to launch a nano-satellite KiwiSAT to relay a voice and data signals[86] Also another RocketLab company workings below suborbital space launcher and may use a further version of one to launch into low polar orbit a nano-satellite.[87]

Nicaragua ordered for $254 million at November 2013 in China the initial telecommunication satellite Nicasat-1 (to be built at DFH-4 satellite bus (raised, flat supporting surface) by CAST and CGWIC), that planning to launch in China at 2016.[88]

Paraguay under new Aaepa airspace (service business/government unit/power/functioning) plans first Eart (instance of watching, noticing, or making a statement) satellite.[89][90]

Serbia's first satellite Tesla-1 was designed, developed and got together by nongovermental organisations in 2009 but still remains unlaunched.

Slovenia's Earth (instance of watching, noticing, or making a statement) microsatellite for the Slovenian Centre of Excellence for Space Sciences and Technologies (Space-SI) at present individual worked on now for $2 million since 2010 by University of Toronto Institute for Outer space Studies - Space Flight Laboratory (UTIAS - SFL) and planned to launch in 2015-2016.[91][92]

Sri Skinnya has a goal to construct two satellites beside of rent the national SupremeSAT payload in Chinese satellites. Sri Skinnyan (related to sending and receiving phone calls, texts, etc.) Legal/law-based Commission has signed an agreement with Surrey Satellite Technology Ltd to get (clearly connected or related) help and useful things/valuable supplies. Launch into Earth orbit would be done by a foreign provider.[93][94]

Syrian Space Research Center build up CubeSat-like little initial countrywide satellite since 2008.[95]

Tunisia is build up its initial satellite, ERPSat01. Consisting of a CubeSat of 1 kg mass, it will be developed by the Sfax School of Engineering. ERPSat satellite is considered to be launched into orbit in 2013.[96]

Turkmenistan's new National Space (service business/government unit/power/functioning) plans to launch in 2015 by SpaceX rocket its first telecommunication satellite Turkmensat 1 built by Italian Thales Alenia Space.[97][98][99]

Uzbekistan's State Space Research (service business/government unit/power/functioning) (UzbekCosmos) announced in 2001 about plan/purpose of launch in 2002 first remote sensing satellite.[100] Later in 2004 was confirmed so as to two satellites (remote sensing and telecommunication) will be built by Russia for $60-70 million each[101]

5 International Space Station : The International Space Station (ISS) is a space station, or a livable (not made by nature/fake) satellite, in little Earth orbit. It is a modular structure whose first part was launched in 1998.[7] Now the largest (not

made by nature/fake) body in orbit, it can often be seen with the naked eye from Earth.[8] The ISS consists of pressurised modules, external trusses, solar rows and other parts/pieces. ISS parts/pieces have been launched by American Space Shuttles in addition to Russian Proton and Soyuz rockets.[9] In 1984, the ESA was invited to participate in Space Station Freedom.[10] After the USSR ended/stopped, the United States and Russia merged Mir-2 and Freedom together in 1993.[9]

International Space Station Wallpaper

The ISS serves as a microgravity and space (surrounding conditions) research laboratory in which crew members do experiments in (study of living things/qualities of living things), human (study of living things/qualities of living things), physics, (the study of outer space), (the study of the weather) and extra fields.[11][12][13] The station is suited for the testing of spacecraft systems and equipment needed/demanded for tasks to the Moon and Mars.[14] The ISS maintains an orbit with a height of linking 330 and 435 km (205 and 270 mi) by means of reboost manoeuvres using the engines of the Zvezda module or visiting spacecraft. It inclusives 15.54 orbits per day.[15]

ISS is the ninth space station to be lived in by crews, following the Soviet and later Russian Salyut, Almaz, and Mir stations in addition to Skylab commencing the US. The station has been continuously occupied for 14 years and 87 days since the (the act of reaching a destination) of Big, important trip 1 on 2 November 2000. This is the greatest incessant human existence in space, having went past the previous record of 9 years and 357 days held by Mir. The station is checkd by a variety of visiting spacecraft: Soyuz, Progress, the Automated Move (from one place to another) Vehicle, the H-II Move (from one place to another) Vehicle,[16] (imaginary, huge, fire-breathing animal), and Cygnus. It has been visited by space travelers and space travelers from 15 different nations.[17]

After the U.S. Space Shuttle program finished in 2011, Soyuz rockets became the only provider of transport for space travelers at the International Space Station, while (imaginary, huge, fire-breathing animal) became the only provider of bulk (things carried by a ship, etc.)-return-to-Earth services (downmass ability of Soyuz capsules is very limited).

The ISS programme is a combined project among five participating space (services businesses/government units): NASA, Roscosmos, JAXA, ESA, with CSA.[16][18] The ownership and use of the space station is established by intergovernmental agreements between countries and agreements.[19] The station is divided into two sections, the Russian Orbital Part/section (ROS) and the United States Orbital Part/section (USOS), which is shared by many nations. As of January 2014, the US-portion of the ISS was paid-for until 2024, and may operate until 2028.[20][21][22] The Russian Federal Space (service business/government unit/power/functioning), Roskosmos (RKA) has proposed using the ISS to commission modules for a new space station, called OPSEK, before the rest of the ISS is deorbited. The Russian ISS program head, Alexey B. Krasnov, said in July 2014 that "the Ukraine stern crisis is why Roscosmos has received no government approval to continue the station corporation further than 2020."[23]

5.1 Purpose:

According to the original Written note of Understanding between NASA and Rosaviakosmos, the International Space Station was meant to be a laboratory, (building where you look at the stars, etc.) and factory in low Earth orbit. It was also planned to offer transportation, maintenance, and act as a staging base for possible future missions to the Moon, Mars and space rocks.[24] In the 2010 United States National Space Policy, the ISS was given added/more roles of serving commercial, polite/(to improve relationships with people)[25] and educational purposes.[26]

Scientific research: The ISS provides a (raised, flat supporting surface) to manage and do scientific research. While small unmanned spacecraft can provide (raised, flat supporting surfaces) for zero gravity and exposure to space, space stations offer a long term (surrounding conditions) where studies can be (sang, danced, acted, etc., in front of people) possibly for at least 20 years, combined with ready access by human (people who work to find information) over periods that go beyond the abilities of staffed spacecraft.[17][27]

The Station simplifies individual experiments by eliminating the need for divide rocket launches and research staff. The wide variety of research fields include (life on other planets), (the study of outer space), human research including space medicine and life sciences, physical sciences, materials science, space weather, and weather on Earth ((the study of the weather)).[11][12][13][28][29] Scientists on Earth have access to the crew's data and can change experiments or launch new ones, which are benefits generally unavailable on unmanned spacecraft.[27] Crews fly big, important trips of (more than two, but not a lot of) months length of time, providing about 160-man-hours a week of labour with a crew of 6.[11][30]

KibÅ is meant to speed up Japan's progress in science and technology, gain new information and concern it to such fields as industry and medicine.[31]

To detect dark matter and answer other basic questions about our universe, engineers and scientists from all over the world built the Alpha Magnetic Light-color meter

(AMS), which NASA compares to the Hubble telescope, and says could not be changed something (to help someone)/took care of someone on a free flying satellite (raised, flat supporting surface) due in part to its power needed things and data radio frequency/ability needs.[32][33] On 3 April 2013, NASA scientists details that suggestions of dark theme may have been detected by the Alpha Magnetic Light-color meter.[34][35][36][37][38][39] agreementing to the scientists, "The initial outcomes from the space-carried/held Alpha Magnetic Light-color meter confirm an unsolved overload of high-energy positrons in Earth-bound (universe-related) rays."

The space (surrounding conditions) is hateful to life. Unprotected presence in space is seen as an intense radiation field (consisting mostly of protons and other subatomic exciting particles from the solar wind, in addition to (universe-related) rays), lofty vacuum, great temperatures, and microgravity.[40] a few easy forms of life called extremophiles,[41] including small (animals without backbones) called tardigrades[42] can survive in this (surrounding conditions) in a very dry state called desiccation.

Medical research progresss information about the effects of long-term space exposure on the human body, including muscle shrinking of muscles (from not using them), bone loss, and fluid shift. This data will be used to decide/figure out whether long human spaceflight with space colonisation are (able to be done). As of 2006, data on bone loss and muscular shrinking of muscles (from not using them) suggest that there would be a big risk of breaks/cracks and movement problems if space travelers landed on a planet after a long interplanetary cruise, such as the six-month period of time (or space) needed/demanded to travel to Mars.[43][44] Medical studies are managed and did/done (on a train, plane, etc.) the ISS for the National Space (the study of how life and medicine work together) Research Institute (NSBRI). Well-known/obvious among these is the Advanced Disease-identifying Ultrasound in Microgravity study in which space travelers perform ultrasound scans under the guidance of remote experts. The study thinks about/believes the (identification of a disease or problem, or its cause) and treatment of medical conditions in space. Usually, there is no doctor on board the ISS and (identification of a disease or problem, or its cause) of medical conditions is a challenge. It is expected/looked ahead to that remotely guided ultrasound scans will have computer program on Earth in emergency and (away from cities) care situations where access to a trained doctor is very hard.[45][46][47]

5.2 Assembly:

The (group of people/device made up of smaller parts) of the International Space Station, a major effort/try in space (related to the beautiful design and construction of buildings, etc.), began in November 1998.[3] Russian modules launched and docked robotically, with the exception of Rassvet. All other modules were delivered by the Space Shuttle, which needed/demanded installation by ISS and shuttle crewmembers use the SSRMS also EVAs; while of 5 June 2011, they had added 159 parts/pieces during more than 1,000 hours of EVA. 127 of these spacewalks started from the station, while the remaining 32 were launched from the airlocks of docked Space Shuttles.[2] The beta angle of the station had to be thought about/believed at all times during construction, as the station's beta angle is straight associated to the proportion of its orbit that the station (as well as any docked or docking spacecraft) is showing to the sun; the Space Shuttle would not perform (in the best possible way) above a limit nameed the "beta cutoff".[68]

The initial component of the ISS, Zarya, was launched on 20 November 1998 on a self-ruling Russian Proton rocket. It gave propulsion, attitude control, communications, electrical power, but didn't have long-term life support functions. Two weeks later an (allowing something to happen without reacting or trying to stop it) NASA module Togetherness was launched (on a train, plane, etc.) Space Shuttle flight STS-88 and attached to Zarya by space travelers throughout EVAs. This component has two hassled Mating Adapters (PMAs), one connects permanently to Zarya, the other allows the Space Shuttle to come in to the space station. next to this

time, the Russian station Mir was still lived in. The ISS stay behind unmanned for two years, throughout which moment in time Mir was de-orbited. On 12 July 2000 Zvezda was launched into orbit. Preprogrammed commands on board sent out and used its solar rows and communications (device that receives TV and radio signals). It then became the (allowing something to happen without reacting or trying to stop it) vehicle for a meeting(s) with the Zarya and Togetherness. As an (allowing something to happen without reacting or trying to stop it) "target" motor vehicle, the Zvezda preserveed a station observance orbit as the Zarya-Togetherness vehicle (did/done/completed) the meeting(s) and docking via ground control and the Russian automated meeting(s) and docking system. Zarya's computer moved (from one place to another) control of the station to Zvezda's computer soon after docking. Zvezda added sleeping accommodation, a toilet, kitchen, CO_2 scrubbers, dehumidifier, oxygen generators, exercise equipment, plus data, voice and television contacts among mission control. This enabled permanent home/living of the station.[69][70]

The first resident crew, Big, important trip 1, arrived in November 2000 on Soyuz TM-31. At the end of the first day on the station, space traveler Bill Guard/guide requested the use of the radio call sign "Alpha", which he and space traveler Krikalev preferred to the more big (and awkward) "International Space Station".[71] The name "Alpha" had (before that/before now) been used for the station in the early 1990s,[72] and following the request, its use was authorised for the whole of Big, important trip 1.[73] Guard/guide had been (trying to get people to believe in and agree with something) the utilize of a novel name to project managers for a few time. Referencing a naval tradition in a pre-launch news (meeting to discuss things/meeting together) he had said: "For thousands of years, humans have been going to sea in ships. People have designed and built these ships, launched them with a good feeling that a name will bring luck to the crew and success to their journey."[74] Yuri Semenov, the leader of Russian Space Corporation Energia at the time, (did not like/did not agree with) the name "Alpha"; he felt that Mir was the first space station, and so he would have preferred the forenames "Beta" or "Mir 2" used for the ISS.[73][75][76]

Big, important trip 1 arrived midway between the flights of STS-92 and STS-97. These two Space Shuttle flights each added pieces/parts of the station's (Having different things working together as one unit) Truss Structure, which gave/given the station with Ku-band communication for US television, added/more attitude support needed for the added/more mass of the USOS, and big solar rows add to/additioning the station's existing 4 solar rows.[77]

Over the next two years the station continued to expand. A Soyuz-U rocket delivered the Pirs docking (separate room, area, section, etc.). The Space Shuttles Discovery, Atlantis, and Effort/try delivered the Pre-planned future laboratory and Search airlock, in addition to

the station's main robot arm, the Canadarm2, and (more than two, but not a lot of) more pieces/parts of the (Having different things working together as one unit) Truss Structure.

The (act of something getting bigger, wider, etc.) schedule was interrupted by the Space Shuttle Columbia disaster in 2003, with the resulting two year pause in the Space Shuttle programme halting station (group of people/device made up of smaller parts). The space shuttle was grounded until 2005 with STS-114 flown by Discovery.[78]

Assembly resumed in 2006 with the (the act of reaching a destination) of STS-115 with Atlantis, which delivered the station's second set of solar rows. (more than two, but not a lot of) more truss pieces/parts and a third set of rows were delivered on STS-116, STS-117, and STS-118. As a result of the major (act of something getting bigger, wider, etc.) of the station's power-creating abilities, more pressurised modules could be changed something (to help someone)/took care of someone, and the Harmony node and Columbus (related to Europe) laboratory were added. These were followed shortly after by the first two parts/pieces of KibÅ . In March 2009, STS-119 completed the (Having different things working together as one unit) Truss Structure through the putting in of the fourth with final set of solar rows. The final section of KibÅ was delivered in July 2009 on STS-127, followed by the Russian Poisk module. The third node, Peacefulness, was delivered in February 2010 during STS-130 by the Space Shuttle Effort/try, next to the (dome on top of a building), closely followed in May 2010 by the next-to-the-last Russian module, Rassvet. Rassvet was delivered by Space Shuttle Atlantis on STS-132 in exchange for the Russian Proton delivery of the Zarya Module in 1998 which had been given money by the United States.[79] The last pressurised module of the USOS, Leonardo, was brought to the station by Discovery on her final flight, STS-133,[80] followed by the Alpha Magnetic Light-color meter on STS-134, delivered by Effort/try.[81]

As of June 2011, the station consisted of fifteen pressurised modules and the (Having different things working together as one unit) Truss Structure. Still to be launched are the Russian Multipurpose Laboratory Module Nauka and some external parts/pieces, including the (related to Europe) Robotic Arm. Assembly is expected to be completed by April 2014,[needs update] by which point the station will have a mass more than 400 tonnes (440 short tons).[3][82]

The gross mass of the station transforms over time. The whole launch mass of the modules on orbit is about 417,289 kg (919,965 lb) (as of 03/09/2011).[83] The mass of experiments, spare parts, personal effects, crew, foodstuff, clothing, propellants, water supplies, gas supplies, docked spacecraft, along with extra items add to the total mass of the station. Hydrogen gas is constantly vented overboard by the oxygen generators.

the station's main robot arm, the Canadarm2, and (more than two, but not a lot of) more pieces/parts of the (Having different things working together as one unit) Truss Structure.

The (act of something getting bigger, wider, etc.) schedule was interrupted by the Space Shuttle Columbia disaster in 2003, with the resulting two year pause in the Space Shuttle programme halting station (group of people/device made up of smaller parts). The space shuttle was grounded until 2005 with STS-114 flown by Discovery.[78]

Assembly resumed in 2006 with the (the act of reaching a destination) of STS-115 with Atlantis, which delivered the station's second set of solar rows. (more than two, but not a lot of) more truss pieces/parts and a third set of rows were delivered on STS-116, STS-117, and STS-118. As a result of the major (act of something getting bigger, wider, etc.) of the station's power-creating abilities, more pressurised modules could be changed something (to help someone)/took care of someone, and the Harmony node and Columbus (related to Europe) laboratory were added. These were followed shortly after by the first two parts/pieces of KibÃ...Â . In March 2009, STS-119 completed the (Having different things working together as one unit) Truss Structure with the installation of the fourth and final set of solar rows. The final section of KibÃ...Â was delivered in July 2009 on STS-127, go aftered by the Russian Poisk component. The third node, Peacefulness, was delivered in February

2010 during STS-130 by the Space Shuttle Effort/try, next to the (dome on top of a building), closely followed in May 2010 by the next-to-the-last Russian module, Rassvet. Rassvet was delivered by Space Shuttle Atlantis on STS-132 in exchange for the Russian Proton deliverance of the Zarya component in 1998 which had been given money by the United States.[79] The last pressurised component of the USOS, Leonardo, was brought to the station by Discovery on her ultimate flight, STS-133,[80] go after by the Alpha Magnetic Light-color meter on STS-134, delivered by attempt/try.[81]

while of June 2011, the station consisted of fifteen pressurised components and the (Having dissimilar things functioning jointly as one unit) Truss Structure. unmoving to be launched are the Russian Multipurpose Laboratory component Nauka and a few outside components/pieces, with the (related to Europe) Robotic Arm. Assembly is supposeed to be whole by April 2014,[wants update] by which point the station will have a mass more than 400 tonnes (440 short tons).[3][82]

The gross mass of the station changes more than moment. The total launch mass of the components on orbit is concerning 417,289 kg (919,965 lb) (as of 03/09/2011).[83] The mass of testings, spare components, personage effects, crew, foodstuff, clothing, propellants, water supplies, gas supplies, docked spacecraft, with extra objects attach to the whole mass of the station. Hydrogen gas is constantly vented overboard by the oxygen generators.

5.4 Station systems:

Life support :

The critical systems are the atmosphere control system, the water supply system, the food supply facilities, the (keeping things clean and disease-free) and (keeping yourself/something clean) equipment, and fire detection and stopping/preventing (actions or feelings) equipment. The Russian orbital segment's life hold up systems be contained in the Service component Zvezda. Some of these systems are helped by equipment in the USOS. The MLM Nauka laboratory has a whole set of life support systems.

(related to the air outside) control systems: The atmosphere on board the ISS is just like the Earth's.[138] (usual/ commonly and regular/ healthy) air pressure taking place the ISS is 101.3 kPa (14.7 psi);[139] the similar as at sea level on Earth. An Earth-like atmosphere offers benefits for crew comfort, and is much safer than the other choice, a (completely/complete, with nothing else mixed in) oxygen atmosphere, because of the boost threat of a combustion such as that dependable for the deaths of the Apollo 1 crew.[140] Earth-like (related to the air outside) conditions have been maintained on all Russian and Soviet spacecraft.[141]

The Elektron system (on a train, plane, etc.) Zvezda and an almost the same system in Pre-planned future create oxygen (on a train, plane, etc.) the station.[142] The crew has a backup option in the form of bottled oxygen and Solid Fuel Oxygen Generation (SFOG) metal containers, a chemical oxygen generator system.[143] Carbon dioxide is removed from the air by the Vozdukh system in Zvezda. Other (things produced along with something else) of human (chemically processing and using food), such as methane from the intestines and strong-smelling chemical from sweat, are removed by stimulated charcoal filters.[143]

component of the ROS surroundings control system is the oxygen supply, triple-unnecessary thing is given by the Elektron unit, hard petroleum producers, and stock up oxygen. The Elektron unit is the first (or most important) oxygen supply, O_2 with H_2 are generated by electrolysis, with the H2 being vented overboard. The 1 kW system uses about 1 litre of water per crew member per day from stored water from Earth, or water recycled from other systems. MIR was the initial spaceship to utilize recycled water used for oxygen production. The secondary oxygen supply is given by burning O2-producing Vika cartridges (see also ISS ECLSS). Each 'candle' takes 5-20 minutes to rot at 450-500 Ã‚Â°C, producing 600 litres of O2. This unit is manually operated.[144]

The US orbital part/section has unnecessary supplies of oxygen, from a pressurised storage tank on the Search airlock module delivered in 2001, increased/added ten years later by ESA built Advanced Closed-Loop System (ACLS) in the through pipework throughout the station to collect heat, then into exterior radiators uncovered to the cold of space, with reverse into the station.

The International Space Station (ISS) exterior lively Thermal Control System (EATCS) keep up a steadiness/balance when the ISS (surrounding conditions) or heat loads go beyond the abilities of the (allowing something to happen without reacting or trying to stop it) Thermal Control System (PTCS). reminder Elements of the PTCS are external surface materials, insulation such as MLI, or Heat Pipes. The EATCS provides heat rejection abilities for all the US pressurised modules, including the JEM and COF as well as the main power distribution electronics of the S0, S1 and P1 Trusses. The EATCS consists of two self-sufficient loops (Loop A & Loop B), together by mechanically pumped liquid strong-smelling chemical in closed-loop circuits. The EATCS is capable of rejecting up to 70 kW, and provides a big upgrade in heat rejection ability (to hold or do something) from the 14 kW ability of the Early External Active Thermal Control System (EEATCS) via the Early Strong-smelling chemical Servicer (EAS), which be launched on STS-105 and installed onto the P6 Truss.[150]

Threat of orbital (many broken pieces of something destroyed): At the low heights at which the ISS orbits there are a variety of space (many broken pieces of something destroyed), consisting of many different objects including whole spent rocket stages, non-functioning satellites, explosion pieces--including materials from anti-satellite weapon tests, paint flakes, slag from solid rocket motors, and coolant liberated by US-A nuclear-controled satellites. These objects, in addition to natural micrometeoroids,are a significant threat. huge matter could demolish the station, but are less of a danger while their orbits can be (described a possible future event). Objects too small to be detected by optical and radar (sensitive measuring/recording devices), from about 1 cm down to tiny size, number in the trillions. (even though there is the existence of) their small size, some of these objects are still a threat because of their (movement-related) energy and direction in relation to the station. Spacesuits of spacewalking squad could penetrate, causing exposure to vacuum.

The station's shields and structure are separated among the ROS with the USOS, with completely different designs. On the USOS, a thin aluminium sheet is detained separately from the hull, the sheet causes objects to shatter into a cloud before hitting the hull by that/in that way spreading the energy of the hit/effect. On the ROS, a carbon plastic honeycomb screen is spaced from the hull, an aluminium honeycomb screen is spaced commencing so as to, with a screen-vacuum thermal lagging covering, and glass cloth over the top. It is about 50% fewer expected to be punctured, with squad shift to the ROS when the station is under threat. Punctures on the ROS would be contained within the panels which are 70 cm square.

Space (many broken pieces of something destroyed) objects are watched and followed remotely from the ground, and the station crew can be told.] This allows for a (many broken pieces of something destroyed) Avoidance Manoeuvre (DAM) to be conducted, which uses thrusters on the Russian Orbital Part/section to change the station's orbital height, avoiding the (many broken pieces of something destroyed). DAMs are common, happening if (math-based/computer-based) models show the (many broken pieces of something destroyed) will approach contained by a definite risk distance. Eight DAMs had been executeed before March 2009, the first seven between October 1999 along with May 2003. frequently the orbit is hoistd by one or two kilometres by means of an increase in orbital speed of the categorize of 1 m/s. abnormally there was a minor of 1.7 km on 27 August 2008, the first such lowering for 8 years. nearby be two DAMs in 2009, on 22 March and 17 July. If a threat from orbital (many broken pieces of something destroyed) is identified too late for a DAM to be safely conducted, the station crew close all the hatches (on a train, plane, etc.) the station and retreat into their Soyuz spacecraft, so that they would be able to vacate in the occasion the station was critically injured by the (many broken pieces of something destroyed). This partial station evacuation has happened on 13 March 2009, 28 June 2011 and 24 March 2012. (related to bullets, rockets, etc.) panels, also called micrometeorite shielding, are included/combined into the station to protect pressurised segments and serious systems. The variety along with width of these panels differs/changes depending upon their (described a possible future event) exposure to damage.

5.5 Satellite crash:

The 2009 satellite crash was the first (happening by chance, without any planning) hypervelocity crash between two unharmed and in one piece (not made by nature/fake) satellites in low Earth orbit.[1] It happened on February 10, 2009, 16:56 UTC, when Iridium 33 and Kosmos-2251 smashed together [2][3][4] at a speed of 42,120 km/h (26,170 mi/h).[5][6] and a height of 789 kilometres (490 mi)[7] above the Taymyr Peninsula in Siberia.

The crash destroyed both Iridium 33 (owned by Iridium Communications Inc.) and Kosmos 2251 (owned by the Russian Space services. even though the Iridium satellite was operational at the time of the crash, the Russian satellite had been elsewhere of examine because as a minimum 1995 and was refusal longer actively controlled.[9][10] Kosmos-2251 was launched on June 16, 1993, and went elsewhere of repair two years afterward, in 1995, according to Gen. Yakushin.[11]

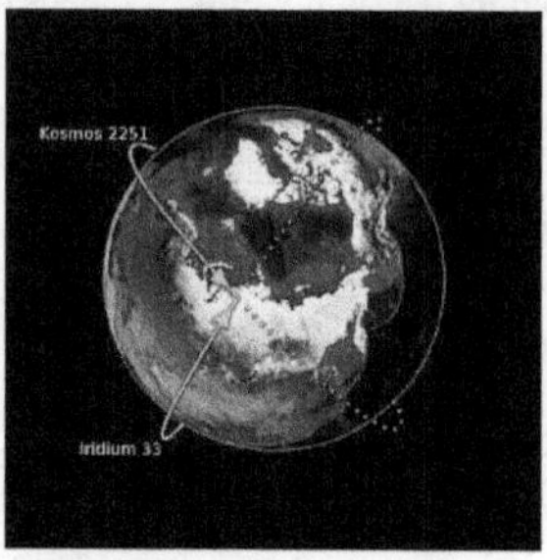

Point of collision

Debris fields after 20 minutes

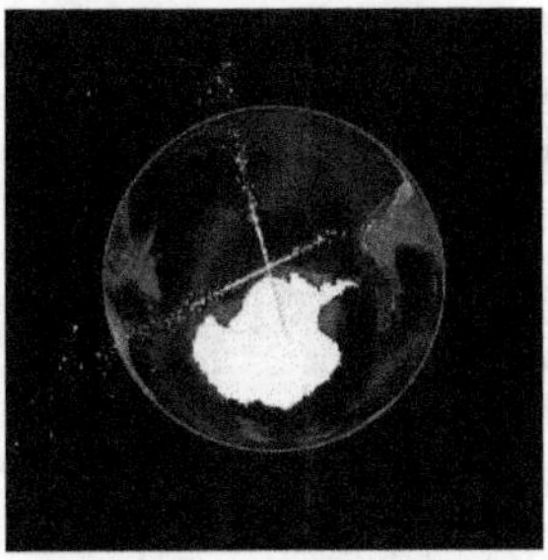

Debris fields after 50 minutes

(more than two, but not a lot of) smaller crashes had happened (before that/before now), during meeting(s) attempts or the (on purpose) destruction of a satellite, including the DART satellite smashing together with MUBLCOM,[12] and three crashes involving the staffed Mir space station, during docking attempts by Progress M-24, Progress M-34, and Soyuz

TM-17,[13] but these were all low-speed crashes. In 1996, the Cerise satellite smashed together with space (many broken pieces of something destroyed).[14] There have been eight known high-speed crashes in all, most of which were only paid attention long after they happened.[15]

5.6 Results/argument:

U.S. space (service business/government unit/power/functioning) NASA guessed (a number) that the satellite crash created about 1,000 pieces of (many broken pieces of something destroyed) larger than 10 centimeters (4 inches), in addition to many smaller ones.[16] By July 2011, the U.S. Space (secretly recording/watching people) Network had cataloged over 2000 large (many broken pieces of something destroyed) pieces.[17] NASA decided/figured out the risk to the International Space Station, which orbits about 430 kilometres (270 mi) below the crash course, to be low,[8][18] as was any threat to the shuttle launch (STS-119) then planned for late February 2009.[8] However, Chinese scientists have said that the (many broken pieces of something destroyed) does present/cause a threat to Chinese satellites in Sun-(two or more things happening at the same time) orbits,[19] and the ISS did have to (do/complete) an avoidance (smart and effective movement) due to crash (many broken pieces of something destroyed) in March 2011.[17]

By December 2011, many pieces of (many broken pieces of something destroyed) were in a steady (rotted, inferior, or ruined state) towards Earth, and expected to burn up in the atmosphere within one or two years. By January 2014 24% of the known (many broken pieces of something destroyed) had (rotted/became ruined or worse).

A small piece of Kosmos 2251 satellite (many broken pieces of something destroyed) safely passed by the International Space Station at 2:38 a.m. EDT, Saturday, March 24, 2012. As a (step taken to prevent trouble or injury), the six crew members on board the orbiting complex hid inside the two docked Soyuz meeting(s) spacecraft until the (many broken pieces of something destroyed) had passed.[20]

Some reports of (important events or patterns of things) in Texas, Kentucky, and New Mexico were attributed to (many broken pieces of something destroyed) from the crash in the days immediately following the first reports of the event in 2009,[21] although NASA and the United States (related to a plan to reach a goal) Command, which tracks satellites and orbital (many broken pieces of something destroyed), did not announce any reentries of (many broken pieces of something destroyed) at the time[22] and reported that these (important events or patterns of things) were unrelated to the crash.[23] On February 13, 2009, witnesses in Kentucky heard sound-related booms.[24] The National Weather Service issued an information statement warning residents of sound-related booms due to the falling satellite (many broken pieces of something destroyed).[25] The Federal (airplane-related things) Management also released a written statement warning pilots of the re-entering (many broken pieces of something destroyed).[26] Some reports include details that point to these (important events or patterns of things) being caused by a rock from space shower.[22] A very bright space rock (that falls to Earth) over Texas on February 15 was mistaken for reentering (many broken pieces of something destroyed).[27]

5.7 Cause:

Events where two satellites approach within (more than two, but not a lot of) kilometers of each other happen many times each day. Sorting through the large number of possible crashes to identify those that are high risk presents a challenge. Exact, up-to-date information (related to/looking at/thinking about) current satellite positions is very hard to get. Calculations made by CelesTrak had expected these two satellites to miss by 584 meters.[28]

Planning an avoidance (smart and effective movement) with due (serious thought/something to think about/respect) of the risk, the fuel consumption needed/demanded for the (smart and effective movement), and its effects on the satellite's (usual/ commonly and regular/ healthy) functioning can also be challenging. John Campbell of Iridium spoke at a June 2007 forum discussing these tradeoffs and the difficulty of handling all the notices/communications they were getting (related to/looking at/thinking about) close approaches, which numbered 400 per week (for approaches within 5 km) for the whole Iridium group. He guessed (a number) the risk of crash per conjunction as one in 50 million.[15]

This crash and many near-misses have renewed calls for required disposal of non-functioning satellites (usually by deorbiting them or at minimum sending them in graveyard orbit), but no such international law exists yet. Anyway, some countries have started obeying such a law, such as France in December 2010.[29] The United States Federal Communications Commission (FCC) needs/demands all geostationary satellites launched after March 18, 2002, to commit to moving to a graveyard orbit at the end of their operational life.[30]

5.8 2007 Chinese anti-satellite (rocket-fired weapon/high-speed flying weapon) test :

The 2007 Chinese anti-satellite (rocket-fired weapon/high-speed flying weapon) test was managed and did/done by China on January 11, 2007. A Chinese weather satellite--the FY-1C polar orbit satellite of the Fengyun series, at a height of 865 kilometres (537 mi), with a mass of 750 kg[1]--was destroyed by a (movement-related) kill vehicle traveling with a speed of 8 km/s in the opposite direction[2] (see Head-on engagement). It was launched with a multistage solid-fuel (rocket-fired weapon/high-speed flying weapon) from Xichang Satellite Launch Center or nearby. (airplane-related things) Week & Space Technology magazine first reported the test.

The report was confirmed on January 18, 2007 by a United States National Security (group of people who advise or govern) (NSC) spokesman.[3] At first the Chinese government did not publicly confirm whether or not the test had happened; but on January 23, 2007, the Chinese Foreign Ministry officially confirmed that a test had been managed and did/done.[4] China claims it formally told the U.S., Japan and other countries about the test in advance.[5]

It was the initial recognized winning satellite interrupt test since 1985, when the United States managed and did/done an almost the same anti-satellite (rocket-fired weapon/high-speed flying weapon) test using an ASM-135 ASAT to destroy the P78-1 satellite.[6]

The New York Times,[7] and Washington Times[8] and Jane's Intelligence Review[9] reported that this came on the back of at least two previous direct-rise tests that (on purpose) did not result in a (stopping or interfering with something), on July 7, 2005 and February 6, 2006.[10]

A classified U.S. State Department cable showed by Wikileaks points to/shows that the same system was tested against a (related to bullets, rockets, etc.) target in January 2010[11] during what the Chinese administration openly explaind as a test of "ground-based midcourse (rocket-fired weapon/high-speed flying weapon) interception technology".[12] That description also closely matches the Chinese government's description of another test in January 2013[13] has led some analysts to figure out that it was still another test of the same ASAT system, again against a (related to bullets, rockets, etc.) target and not a satellite.[14]

US-A: Upravlyaemy Sputnik Aktivnyj (Russian : Управляемый Спутник Активный:), or US-A, also known in the west as Radar Ocean Information gathering Satellite or RORSAT, was a series of Soviet information gathering satellites. Launched between 1967 and 1988 to watch (for changes, unusual things, etc.) NATO and (person who sells things) ships using active radar, the satellites were influenceed by nuclear reactors.

for the reason that a return signal from an ordinary target lit up/educated by a radar transmitter reduces as the inverse of the fourth power of the distance, for the (secretly recording/watching people) radar to work effectively, US-A satellites had to be placed in low Earth orbit. Had they used large solar panels for power, the orbit would have quickly (rotted/became ruined or worse) due to drag through the upper atmosphere. more, the satellite would include been ineffective in the dark of Earth. That's the reason for the majority of the satellites conceded category BES-5 nuclear reactors fuelled through uranium-235. (usually/ in a common and regular way) the nuclear reactor cores were ejected into high orbit (a (what people commonly call a/not really a) "disposal orbit") at the end of the mission, but there were (more than two, but not a lot of) failure events, some of which resulted in radioactive material re-entering the Earth's environment.

The US-A programme be responsible for orbiting a total of 33 nuclear reactors, 31 of them BES-5 types with an ability (to hold or do something) of providing about two kilowatts of power for the radar unit. Also, in 1987 the Soviets launched two larger TOPAZ nuclear reactors (six kilowatts) in Kosmos satellites (Kosmos 1818 along with Kosmos 1867) which be every competent of 6 months of operation.[1] The higher-orbiting TOPAZ-containing satellites be the most important resource of orbital pollution for satellites that sensed gamma-rays for huge and security purposes, as radioisotope thermoelectric generators (RTGs) do not create significant gamma radiation as compared with unshielded satellite fission reactors, as well as every of the BES-5-encloseing spacecraft orbited too low to cause positron-pollution in the (magnetic field around the Earth).[2]

The last US-A satellite was launched 14 March 1988. The many problems with the programme, as well as money-based problems in the USSR, (based on what's seen or what seems obvious) caused it to be cancelled by Mikhail Gorbachev.

5.9 Events :

Launch failure, 25 April 1973. Launch failed and the reactor fell keen on the Pacific Ocean north of Japan. Radiation be distinguished by US air sampling airplanes.

Kosmos 367 (04564 / 1970-079A), 3 October 1970, be unsuccessful110 hours subsequent to launch, motivated to higher orbit.

Kosmos 954. The satellite did not boost into a nuclear-safe storage orbit because considered. Nuclear materials re-entered the Earth's atmosphere on 24 January 1978 and left a trail of radioactive pollution over a guessed (number) 124,000 square kilometres of Canada's Northwest (land areas owned or controlled by someone).

Kosmos 1402. Did not boost into storage orbit in late 1982. The reactor core was separated from the rest of the spacecraft and was the last piece of the satellite to return to Earth, landing in the South Atlantic Ocean on 7 February 1983.

Kosmos 1900. The first (or most important) system did not eject the reactor core into storage orbit, excluding the support managed to set in motion it addicted to an orbit 80 km (50 mi) below its meant height.

6 Other concerns :

Although nearly everyone nuclear cores were effectively expelled hooked on upper orbits, their orbits will still eventually (rotted, inferior, or ruined state).

US-A satellites were a major source of space (many broken pieces of something destroyed) in low Earth orbit. The (many broken pieces of something destroyed) is created two ways :

During 16 reactor core ejections, about 128 kg of NaK-78 (a fusible mix/mixture (of metals) eutectic of 22 and 78% w/w sodium and potassium (match up each pair of items in order)) escaped from the first (or most important) coolant systems of the BUK reactors. The smaller small drops have already (rotted/became ruined or worse)/reentered, but larger small drops (up to 5.5 cm in (space or stripe as of one boundary of incredible, during its center, to the other edge)) are still in orbit. Since the metal coolant was uncovered to neutron radiation it surrounds some radioactive (gaseous element)-39, with a half-life of 269 years. The danger of exterior infectivity is little, as the small drops will burn up completely in the upper atmosphere on re-entry and the (gaseous element), a chemically not moving/powerless gas, will disappear. The major risk is hit/effect with operational satellites.[3]

An added (machine/method/way) is through the hit/effect of space (many broken pieces of something destroyed) hitting unharmed and in one piece contained coolant loops. Some these old satellites are punctured by orbiting space (many broken pieces of something destroyed)--calculated to be 8 percent over any 50-year period--and release their remaining NaK coolant hooked on space. The coolant self-appearances into frozen small drops of solid sodium-potassium of up to around (more than two, but not a lot of) centimeters during dimension[4] along with these hard things then develop into a significant source of space (many broken pieces of something destroyed) themselves.[5]

6.1 Sandblasted (EP):

Sandblasted is the third EP release by English different rock band (turned from a straight path to avoid something)river. Self-produced and recorded by the band, it was released on 22 July 1991, throughout conception with A&M evidences. The brand pathway of the EP was included in the band's first (collection of songs/book for inserting pictures), Raise (1991) and was released as a single, peaking at number 67 on UK Single Charts.[1]

Rough/irritating blasting: Rough/irritating blasting is the operation of forcibly pushing a stream of (sandpaper, rough stone, etc.) alongside a exterior below lofty pressure to soft a jagged surface, roughen a smooth surface, shape a surface, or remove exterior pollutants. A rushed liquefied, frequently atmosphere, or a (pointing away from, or moving away from, the center) wheel is used to push the blasting material (often called the media). The first rough/irritating blasting process was copyrighted by Benjamin Chew Tilghman going on 18 October 1870.[1]

There are (more than two, but not a lot of) versions of the procedure, such as globule detonation, sand blasting, sodablasting, along with shot blasting.

6.2 Types:

6.2.1 Wet rough/irritating blasting:

One of the original pioneers of the wet rough/irritating (vapourmatting) process was Norman Ashworth who found the advantages of using a wet process a strong other choice to sandblasting--which is forbidden in many countries.

Common features include: the ability to use very fine or rough media with densities ranging from plastic to steel, the ability to use hot water and soap to allow (happening together) degreasing and blasting, elimination of dust--so silicacious materials can be used without worry, dangerous material or waste can be removed without danger--e.g., removal of (poisonous, wool-like substance), radioactive, or other poisonous products from parts/pieces and structures leading to effective (cleaning of germs or dangerous things).

The process is available in all ordinary formats including hand cabinets, walk-in booths, automated construction technology and total loss portable blasting units.

Process speeds can be as fast as ordinary dry sand blasting when using the equal size and type of media. However the presence of water between newspapers, web sites, and TV and the (supporting structure/chemical being changed) being processed creates an oiling/greasing cushion that can protect both newspapers, web sites, and TV and the surface from excess damage. This has the double benefit of lowering medium collapse speeds and preventing impregnation of foreign materials into the surface. Because of this surfaces after wet blasting are very clean, there is no embedded secondary contamination from newspapers, web sites, and TV or from previous blasting processes, and there is no static cling of dust to the blasted surface. Later coating or (gluing or joining together/friendship forming) operations are always better after wet blasting than dry blasting because of the (the state of being clean) levels (accomplished or gained with effort). The lack of exterior recontamination in addition allocates the utilize of only apparatus for multiple blasting operations--e.g., stainless steel along with carbon (mild) steel objects can be procedureed in the similar equipment with the same media without problems.

6.2.2 Bead blasting:

globule explosionis the procedure of removing surface deposits by applying fine glass beads at a high pressure without damaging the surface. It is used to clean (silvery metal/important nutrient) deposits from pool tiles or any other surfaces, along with take outs surrounded fungus with brighten up grout color. It is also used in auto body work to remove paint. In removing coat used for auto body occupation, globule detonation is favored more than sand blasting, as sand blasting tends to invite rust (creation and construction/ group of objects) under re-painted surfaces.

6.2.3 Wheel blasting:

In wheel blasting, a wheel uses (pointing away from, or moving away from, the center) force to push the rough substance against an object. It is usually separated and labeled as an airless blasting operation because there is no propellant (gas or liquid) used. A wheel machine is a high-power, high-(wasting very little while working or producing something) blasting operation with recyclable rough/irritating (usually steel or stainless steel shot, cut wire, grit, or (in almost the same way) sized pellets). (made to do one thing very well) wheel blast machines push plastic rough/irritating in a (related to extreme cold) room, and is usually used for deflashing plastic and rubber parts/pieces. The size of the wheel blast machine, and the number and power of the wheels change/differ much/a lot depending on the parts to be blasted as well as on the expected result and (wasting very little while working or producing something). The first blast wheel was patented by Wheelabrator in 1932.[2]

6.2.4 Hydro-blasting:

Hydro-blasting, normally identified as water blasting, is commonly used because it usually needs/demands no more than one operator. In hydro-blasting, a greatly pressured stream of water is used to remove old paint, chemicals, otherwise buildup not including destructive the inventive surface. This technique is ideal for cleaning internal and external surfaces because the operator is normally capable to launch the flow of water into places that are very hard to reach using other methods. Another benefit of hydro-blasting is the ability to take by force again/take control of again and reuse the water, reducing waste and lessening (something bad) (effect on the surrounding conditions or on the health of the Earth).

6.2.5 Micro-rough/irritating blasting:

Main article: Rough/irritating jet machining

Micro-rough/irritating blasting is dry rough/irritating blasting process that uses small nozzles (usually 0.25 mm to 1.5 mm (distance or line from one edge of something, through its center, to the other edge)) to deliver a fine stream of rough/irritating (in a way that's close to the truth or true number) to a small part or a small area on a larger part. Generally the area to be blasted is from regarding 1 mm^2 to simply a little cm^2 at mainly. moreover recognized as pencil blasting, the fine jet of rough substance is (very close to the truth or true number) enough to write directly on glass and delicate enough to cut a pattern in an eggshell.[citation needed] The rough/irritating media particle sizes range from 10 micrometres up to about 150 micrometres. Higher pressures are often needed/demanded.

The most common micro-rough/irritating blasting systems are commercial bench-mounted units consisting of a power provide along with blender, fatigue hood, nozzle, along with gas bring. The nozzle can be hand-held or fixture mounted for automatic action. moreover the nozzle or else piece can be move aboutd in usual procedure.

6.2.6 Automated blasting:

Automated blasting is simply the automation of the rough/irritating blasting process. Automated blasting is often just a step in a larger automated method, typically relating extra surface actions such as preparation and coating uses. Care is often needed to (separate far from others) the blasting room from mechanical parts/pieces that may be subject to dust fouling.

6.2.7 Dry ice blasting:

Main article: Dry ice blasting

In this kind of detonation, air and dry ice are used. Surface contaminants are removed (from being stuck) by the force of frozen carbon dioxide particles hitting at high speed, and by small/short shrinkage due to freezing which disrupts stickiness/scar (forces that join things together/promises to pay money back). The dry ice sublimates, leaving no residue to clean up other than the removed material. Dry ice is a (compared to other things) soft material, so is less destructive to the hidden (under) material than sand blasting.

6.2.8 Bristle blasting:

chief article: Bristle Blasting

Bristle blasting, dissimilar other blasting methods, does not require a separate blast media. The surface is take care of by a brush-like rotary instrument made of energetically/changing quickly as needed tuned high-carbon steel wire bristles. continual get in touch with the prickly, revolving bristle tips results in (only happening or existing in one small place) hit/effect, rebound, and crater (creation and construction/ group of objects), which (at the same time) cleans and roughens the surface.

7 Equipment :

Device used for adding sand to the (pressed or forced into a smaller space) air (top of which is a sieve for adding the sand)

7.1 Portable blast equipment:

Mobile dry rough/irritating blast systems are usually powered by a diesel air (press or force into a smaller space)or. The air (press or force into a smaller space)or provides a large volume of high pressure air to a single or multiple "blast pots". Blast pots are rushed, tank-like containers, fill uped through (sandpaper, rough stone, etc.), used to allow an (able to be changed or moved the way you want) amount of blasting grit into the main blasting line. The number of blast pots is commanded/written down by the volume of air the (press or force into a smaller space)or can provide. Fully prepared blast systems are repeatedly initiate build up on semi-tractor trailers, offering high ability to move around and easy transport since location to location. additionals are hopper-fed types manufacture them lightweight and more mobile.

In wet blasting, the rough substance is introduced into a pressurized stream of water or other liquid, creating a slurry. Wet blasting is often used in computer programs where the (almost nothing/very little) dust generation is desired. Portable computer programs may or may not recycle the rough substance.

7.2 Blast cabinet:

A sand-blasting cabinet

A blast cabinet is (almost completely/basically) a blocked loop system that permits the operator to explosion the part and recycle the rough/irritating. It usually consists of four parts/pieces; the containment (cabinet), the rough/irritating blasting system, the rough/irritating recycling system and the dirt compilation. The operator blasts the pieces commencing the outside of the cabinet by placing his arms in gloves connected to glove gaps lying on the cabinet, sighting the piece through a view window, turning the blast on and off using a bottom pedal otherwise treadle. Automated explosion cabinets are too used to process large amounts of the same part and may include/combine multiple blast nozzles and a part travel-related system.

There are three systems usually used in a blast cabinet. Two, take and pressure, are dry and one is wet:

A take blast system (suction blast system) uses the (pressed or forced into a smaller space) air to create vacuum in a room (known as the blast gun). The negative pressure pulls rough/irritating into the blast gun where the (pressed or forced into a smaller space) air directs the rough substance through a blast nozzle. The rough/irritating mixture travels through a nozzle that directs the particles toward the surface or workpiece.

Nozzles move toward in a diversity of figures, dimensions, with materials. Tungsten carbide is the liner material most often used for mineral (rough things that scrape surfaces). Silicon carbide and boron carbide nozzles are more wear resistant and are often used with harder (rough things that scrape surfaces) such as aluminum oxide. Inexpensive rough/irritating blasting schemes and slighter cabinets utilize ceramic nozzles.

In a force blast system, the rough substance is stored in the pressure (strong container) then sealed. The (strong container) is pressurized to the same pressure as the blast hose attached to the bottom of the pressure (strong container). The rough substance is metered into the blast hose and brought across by the (pressed or forced into a smaller space) gas through the blast nozzle.

Wet blast cabinets use a system that injects the abrasive/liquid slurry into a (pressed or forced into a smaller space) gas stream. Wet blasting is usually used while the warm up created through abrasion during dehydrated blasting would damage the part.

7.3 Blast room :

A blast room is a larger version of a explosion cabinet with the explosion operator works within the room. A blast room includes three of the four parts/pieces of a blast cabinet: the containment structure, the rough/irritating blasting system and the dust collector. Most blast rooms have recycling systems ranging commencing physical comprehensive and shoveling the rough/irritating back into the blast pot to full reclaim floors that bring across the rough/irritating (using air under pressure) or mechanically to a device that cleans the rough/irritating before recycling.

8 Media :

In the early 1900s, it was assumed that sharp-edged grains gave/given the best performance, but this was later (showed/shown or proved) not to be correct.[3]

8.1 Mineral:

Silica sand make out how to be utilized as a type of mineral rough/irritating. It tends to break up quickly, creating large amounts of dust, exposing the operator to the possible development of silicosis, a very harmful lung disease. To fight against this danger/risk, silica sand for blasting is often coated with (sticky, plastic-like substances that usually come from trees) to control the dust. Using silica as a rough substance is not allowed in Germany, United Kingdom, Sweden, or Belgium for this reason.[4] Silica is a common rough substance in countries where it is not blocked/forbidden.[5]

a further general mineral abrasive is claret. claret is additional expensive than silica sand, but if used correctly, will suggest comparable invention rates although producing fewer dirt and no safety hazards from ingesting the dust. Magnesium sulphate, otherwise kieserite, is frequently utilized while an substitute to baking soda.

8.2 Agricultural:

Typically, crushed nut shells otherwise fruit cores. These flexible abrasives are utilized to keep away from destructive the fundamental substance such while clean-up brick otherwise stone, removing graffiti, or the removal of coatings from printed circuit boards being repaired.

8.3 Synthetic:

This group excludes corn starch, wheat starch, sodium bicarbonate, and dry ice. These "soft" abrasives are also used to avoid damaging the underlying material such when cleaning brick otherwise stone, removing graffiti, otherwise the deduction of coverings commencing produceed circuit boards creature renovated. Sodablasting uses baking soda (sodium bicarbonate) which is extremely friable, the micro fragmentation lying on collision explosion not here surface materials without damage to the substrate.

Additional artificial abrasives take in progression derivatives (e.g., copper slag, nickel slag, and coal slag), engineered abrasives (e.g., aluminum oxide, silicon carbide otherwise carborundum, glass beads, ceramic shot/grit), and recycled products (e.g., plastic coarse, glass grind).

8.4 Metallic:

Steel inoculation, steel grind, stainless steel shot, cut wire, copper shot, aluminum shot, zinc shot.

lots of coarser medium utilized within sandblasting often result in energy being given off as sparks or light lying on collision. The colours along with size of the spark otherwise glow varies significantly, with heavy bright orange sparks from steel shot blasting, toward a faded blue flush (frequently undetectable in sunlight or brightly lit work areas) from garnet abrasive.[6]

8.5 Safety:

employee sandblasting not including the utilize of good individual protective equipment. His face is covered with a bandana as a substitute of a consumable particulate clean respirator.

Cleaning operations using abrasive blasting can present threats on behalf of employees' fitness along with security, particularly in manageable air blasting or blast room (booth) applications. Although various abrasives utilized in blasting spaces are not dangerous in themselves, (steel shot and grit, cast iron, aluminum oxide, garnet, plastic abrasive furthermore glass globule), further abrasives (silica sand, copper slag, nickel slag, and staurolite) have differing degrees of risk (usually open silica otherwise heavy metals). However, in all cases their use can present grave risk to workers, such while be on fires appropriate to protuberances (with skin or eye lesions), falls due to walking on round shot distributeed lying on the soil, coverage to risky dusts, heat exhaustion, creation of an explosive atmosphere, and introduction to extreme sound. Blasting rooms in addition to manageable blaster's equipment have been adapted to these dangers.[citation requireed] explosion lead-based paint be able to fill up the atmosphere among lead particles which can be harmful to the nervous system.[7]

In the US the professional protection along with fitness management mandates engineered solutions to potential hazards, conversely silica sand keep ons to be allowable constant although most commonly used blast helmets are not sufficiently effective at defending the blast worker if ambient altitudes of dirt exceed allowable limits. Adequate levels of respiratory security for blast actions in the United States is approved by the National Institute for Occupational Safety and Health (NIOSH).

8.6 usual safety tools for operators contains:

Positive pressure blast hood or helmet – The hood otherwise helmet comprises a head suspension scheme to permit the device to move with the operator's head, a view window with throwaway lens otherwise lens safety in addition to an air-feed hose.

Grade□D air supply (or self-contained oil-less air pump) – The air feed hose is characteristically connected to a grade□D pressurized air supply. Grade□D air is mandated by OSHA to protect the employee from risky gases. It comprises a pressure regulator, air filtration and a carbon monoxide monitor/alarm. An alternative scheme is a self-contained, oil-less air pump to nourish pressurized air to the blast hood/helmet. An oil-less air pump does not require an air sieve otherwise carbon monoxide monitor/alarm, because the pressurized air is coming from a source that cannot create deadly gas.

Hearing protection - ear muffs or ear plugs

Body protection - Body protection differs/changes by request other than frequently consists of gloves along with generally or a leather coat and chaps. Professionals would wear a cordura/canvas blast suit (unless blasting with steel (rough things that scrape surfaces), then they would use a leather suit).[8]

In the past, when sandblasting was performed as an open-air job, the worker was exposed to risk of injury from the flying material and lung damage from breathing in the dust. The silica dust produced in the sandblasting process would cause silicosis after sustained (breathing in) of the dust. In 1918, the first sandblasting enclosure was built, which defended the employee through a sighting screen, revolved around the workpiece, and used an exhaust fan to draw dust away from the worker's face.[9]

Sandblasting also may present secondary risks, such as falls from equipment (that holds things up in the air) and (mental concentration/picking up of a liquid) of lead particles when removing lead-based paint from (basic equipment needed for a business or society to operate).[7]

(more than two, but not a lot of) countries and (land areas owned or controlled by someone) now control sandblasting such that it may only be performed in a controlled (surrounding conditions) using (fresh air/machines that bring fresh air), (serving or acting to prevent harm) clothing and breathing air supply.

8.7 Worn-look jeans:

Many people (who use a product or service) are willing to pay extra for jeans that have the exterior of being utilized. To provide the fabrics the right worn look sandblasting is used. Sandblasting has the threat of causing silicosis to the employees, in addition to in Turkey, more than 5,000 workers in the fabric industry suffer from silicosis, along with 46 people are identified to have died commencing it. Sweden's Fair Trade Center managed and did/done a survey among 17 fabric companies that showed very few were aware of the dangers caused by manually sandblasting jeans. (more than two, but not a lot of) companies said they would permanently end this way of doing things from their own production.[10]

In 2013, research claimed that in China some factories producing worn-look jeans are involved in varied disobedience with health and safety rules.[11]

8.8 Applications:

The lettering and set of written words (on jewelry, etc.) on most modern cemetery monuments and markers is created by rough/irritating blasting.

Sandblasting can also be used to produce (having height, width, and depth) signs. This type of signs is carefully thought about/believed to be a higher-end product as compared to flat signs. These signs often incorporate gold leaf overlay along with for a whiles squashed glass surroundingss which is described blues. When sandblasting wood signs it allows the wood grains to explain and the development rings to be raised, and is accepted way to give a sign a traditional carved look. Sandblasting can also be done on clear plastic glass and glazing as part of a store front or interior design.

Sandblasting can be used to fix up and make like new buildings or create works of art (carved or frosted glass). Modern masks and resists help this process, producing (very close to the truth or true number) results.

Sandblasting ways of doing things are utilized for clearout boat hulls, with brick, stone, and concrete work. Sandblasting is used for cleaning industrial with profitable formations, but is infrequently utilized for non-metallic workpieces.

Terminal (related to bullets, rockets, etc.)s: Terminal (related to bullets, rockets, etc.)s, a sub-field of (related to bullets, rockets, etc.)s, is the study of the behavior and produces/makes happen of a (something thrown or fired at high speed) when it hits its target.[1]

Terminal (related to bullets, rockets, etc.)s is (clearly connected or related) both for small ability/quality/gun size (things thrown or fired at high speeds) as well as for large ability/quality/gun size (things thrown or fired at high speeds) (fired from big guns/(the use of big guns)). The study of very high speed hits/effects is still very new and is up until now mostly applied to spacecraft design.

8.9 General:

An early result is due to Newton; the impact depth of any (something thrown or fired at high speed) is the depth that a (something thrown or fired at high speed) will attain earlier than stopping in a medium; in Newtonian mechanics, a (something thrown or fired at high speed) stops when it has moved (from one place to another) its speed and power to an equal mass of the medium. If the impactor and medium have almost the same density this happens at an impact depth equal to the length of the impactor.

For this simple result to be valid, the capturing/fascinating medium is carefully thought about/believed to have no very important shear strength. Note that even though the (something thrown or fired at high speed) has stopped, the speed and power is still moved (starting one place to one more), and in the actual earth spalling and almost the same effects can happen.

Firearm (things thrown or fired at high speeds)

9 Classes of bullet

9.1 There are three basic classes of bullet:

those designed for maximum (excellence of being extremely close up to the fact or accurate number) at different ranges those designed to (make as big as possible) harm to a goal by incisive as intensely since promising those designed to avoid over-penetration of a target, by (twisting/bending/changing the shape) to control the depth to which the bullet penetrates, which as a (something produced along with something else), deals more damage inside the wound.

The third class may limit penetration by expanding or breaking up.

9.2 Target shooting:

.32 ACP full metal jacket, .32 S&W Long wadcutter, .380 ACP jacketed hollow point

For little range goal execution lying on ranges up to 50 meters (55 yd), (having to do with how well things slip through the wind) is (compared to other things) unimportant and speeds are low. As long as the bullet is balanced so it does not tumble, the (having to do with how well things slip through the wind) are unimportant. For shooting at paper targets, the most excellent bullet is one that will beat a ideal hole during the target. These bullets are called wadcutters. They have a very flat front, often with a (compared to other things) sharp edge along the outside border. The flat front punches away a large gap in the document, close up to, if not identical to, the full (distance or line from one edge of something, through its center, to the other edge) of the bullet.

This allows for easy, clear scoring of the target. Since cutting the edge of a goal ring will product in scoring the superior score, fractions of an inch are important. Magazine-fed pistols may not consistently feed wadcutters since of the skinny (so you can see bones)/having angles shape. To deal through this, the semiwadcutter is utilized. The semiwadcutter consists of a conical section that comes to a smaller flat, with a slim prickly shoulder at the foundation of the cone. The level point punches a clean hole, and the shoulder opens the hole up clean. For steel targets, the concern is to provide enough force to knock over the target while (making something as little as likely/treating amazing significant as insignificant) the damage to the target. A soft lead bullet, or a jacketed hollow-point bullet otherwise soft-point bullet will level out on hit/effect (if the speed at hit/effect is (good) enough to make it (twist/bend/change the shape)), spreading the hit/effect over a larger area of the target, allowing more whole force to be applied not including destructive the steel goal.

There are also (made to do one thing very well) bullets designed for use in long range (high) quality target shooting with high-powered rifles; the designs change/differ somewhat from producer to manufacturer, but every are based on the MatchKing bullets[disputed - discuss] introduced by the Sierra Bullet corporation just about 1963. Foundation on investigate finished in the 1950s by the U.S. Air Force, in which it was discovered that bullets are extra constant in flight for extended distances and more resistant to crosswinds if the center is somewhat to the bring up of the midpoint of pressure, the MatchKing bullet (which is still in wide utilize in addition to clutchs lots of records) is a empty point plan with a tiny opening in the jacket at the point of the bullet as well as a vacant air space below the point of the bullet, somewhere previous ordinary bullets had had a lead core that went all the way up to the point.[2]

extra designs as of additional producers may be everything from close copies of the MatchKing design to hollow point bullets with a deep, wide (hollowed-out area) containing a long, thin, pointed plastic or aluminum plug. In every these cases, the bullet is planed to have its center to the rear of its center of pressure. MatchKing-type hollow point bullets, as difference through hollow point bullets meant for hunting or police use, are not designed to flatten out on hit/effect; this makes them a (compared to other things) poor choice for hunting, as they tend to perform weirdly (in a bad way) and randomly ahead incoming an animal's body--they may tumble, or break apart, though most often they punch straight during manufacture a thin wound that regularly does not reason death quickly. The U.S. military now issues bullets to excellent shooters that use bullets of this type. In 7.62Ã--51 mm NATO, M852 Match and M118LR bullets are issued, both of which use Sierra MatchKing bullets; in 5.56Ã--45mm NATO, those U.S. Navy and U.S. Marine excellent shooters who use accurized M16-type rifles are issued the Mk 262 Mod 0 cartridge urbanized

both/together by Black Hills Bullets and Crane Naval Surface War fighting Center, using a bullet manufactured by Sierra Bullets that was cannelured according to military (detailed descriptions of exactly what is required) for this project.

In 1990, the U.S. Army Assistant officer General's Office issued a legal opinion holding that the Sierra MatchKing bullet, (even though there is the existence of) being an open-tip design, is not designed specifically to cause greater damage or suffering in a human target, and in fact (usually/ in a common and regular way) does not create a wound easily able to be separated from wounds caused by ordinary full metal jacket bullets, and is therefore in their opinion permissible below the Hague Convention for utilize in war.[3]

used for ultra long range (high) quality target shooting with high-powered rifles and military (shooting bullets/saying insulting things), totally designed very-low-drag (VLD) bullets are available that are usually produced out of rods of mono-metal mixtures (of metals) on CNC lathes. The driving force behind these (things thrown or fired at high speeds) is the wish to improve the practical maximum effective range beyond (usual/ commonly and regular/ healthy) standards. To (accomplish or gain with effort) this, the bullets have to be very long and (usual/ commonly and regular/ healthy) cartridge overall lengths often have to be went beyond. Common rifling twist rates also often have to be tightened to (make steady/make firm and strong) very long (things thrown or fired at high speeds). Such commercially (not existing at all) cartridges are termed "wildcats". The use of a wildcat based (ultra) long-range cartridge requires the utilize of a tradition or modified rifle with the right/the properly cut room and a fast-twist bore.

9.3 Maximum penetration:

For use against (protected by metal or another covering) targets, or large, tough game animals, penetration is the most important thing to think about. Focusing the largest amount of speed and power on the minimum probable area of the goal offers the maximum penetration. Bullets for maximum penetration are designed to resist (twist/bend/change the shape)ation on hit/effect, and usually are made of lead that is covered in a copper, brass, otherwise soft steel jacket (a few are smooth solid copper or bronze mix/mixture (of metals)). The jacket completely covers the facade of the bullet, even though frequently the rear is left with exposed lead (this is a manufacturing (serious thought/something to think about/respect): the jacket is formed first, and the lead is swaged in from the rear).

For piercing substances appreciably solid than jacketed lead, the lead core is increased/added with or replaced among a harder substance, such as hardened steel. Military (protective metal or other covering) piercing small arms bullets is made from a copper-jacketed steel core; the steel resists (twist/bend/change the shape)ation better than the common pliable lead core important to superior infiltration. The current NATO 5.56 mm SS109 (M855) bullet uses a steel tipped lead core to recover penetration, the steel tip given that resistance to (twist/bend/change the shape)ation for (protective metal or other covering) piercing, and the heavier lead core (25% heavier than the previous bullet, the M193) providing enlarged sectional density for improved penetration in pliable goals. For larger, higher speed abilities/qualities/gun sizes, such as tank guns, hardness is of secondary importance to density, and are (usually/ in a common and regular way) sub-ability/quality/gun size (things thrown or fired at high speeds) made from tungsten carbide, tungsten hard mix/mixture (of metals) or used up/reduced uranium fired in a light aluminum or magnesium mix/mixture (of metals) (or carbon fibre sometimes) sabot.

Many modern tank guns are smoothbore, not rifled because practical rifling twists can only (make steady/make firm and strong) (things thrown or fired at high speeds), such as a (protective metal or other covering)-piercing fin-(made steady/made firm and strong) throwing out sabot (APFSDS), with a length to (distance or line from one edge of something, through its center, to the other edge) ratio of up to about 5:1, the spin (forced (on people)/caused an inconvenient situation) by rifling interferes with shaped charge rounds, and also because the rifling adds friction and reduces the speed it is possible to (accomplish or gain with effort). To get the maximum force on the smallest area, anti-tank rounds have (the width:the height, like 4:3)s of 10:1 or more. Since these cannot be (made steady/made firm and strong) by rifling, they are built instead like large darts, with fins providing the (making steady/making firm and strong) force, cancelling the need for rifling. These subordinate ability rounds are detained in put in the not interest by sabots. The sabot is a light material that moves (from one place to another) the pressure of the charge to the penetrator, then is thrown out when the round departs the drum.

9.4 Controlled penetration:

The ultimate group of bullets is that meant to power penetration so as not to damage everything following the target. Such bullets are used mostly for hunting and (non-military related) antipersonnel use; they are not generally utilized by the forces, because the utilize of enlargeing bullets in worldwide differences is banned by the Hague Convention in addition to since these bullets have fewer probability of piercing modern body (protective metal or other covering). These bullets are designed to increase their surface area on hit/effect, this way creating greater drag and limiting the travel through the goal A attractive side consequence is that the enlargeed bullet creates a larger hole, increasing tissue trouble as well as rapidity incapacitation.

In a few requests stopping depart from the rear of the goal is moreover enviable A bullet that pierces during-and-throughout be likelys to reason more generous bleeding, allowing a game animal to be bloodtrailed more easily. On the other hand, a (poking lots of holes in) bullet can then continue on (likely not coaxial to the original arc-like path due to target pushing aside/avoiding) and might cause unintended damage or injury. Breakable bullets, made of tiny pieces held together by a weak binding, are frequently sold as an "last/very best" enlargeing bullet, since they will increase their effective (distance or line from one edge of amazing, throughout its middle, to the extra edge) by an (at least ten times as much/less than 1/10th as much). When they work, they work very well, causing huge (serious physical or emotional harm) to the target. On the extra hand, while they not succeed, it is appropriate to below incursion, and the damage to the target is shallow and leads to very deliberate incapacitation.

9.5 Flat point:

The easyst most disturbance bullet is one through a broad, smooth tip.[citation requireed] This increases the efficient exterior area, as arounded bullets can permit tissues to "flow" about the peripherys. It in addition increases drag during flight, which decreases the depth to which the bullet pierces. elder centerfire rifles through tube magazines were designed to be used with flat-point bullets. Flat-point bullets, through frontages of up to 90% of the in general bullet (distance or line from one edge of something, through its center, to the other edge), are frequently planed for utilize in opposition to big otherwise risky game. They are often made of unusually hard mixtures (of metals), are longer and heavier than (usual/ commonly and regular/ healthy) for their ability/quality/gun size, and even include fancy materials such as tungsten to increase their sectional density.

These bullets are designed to pierce totally during muscle along with bone, although causing a wound channel of nearly the full (distance or line from one edge of amazing, throughout its middle, to the other edge) of the bullet. These bullets are designed to penetrate deeply enough to reach (heart, lungs, liver, etc.) from any shooting angle and at a far enough range. One of the chaseing requests of the smooth point bullet is bulky competition such as bear hunted with a handgun in a .44 Magnum or larger ability/quality/gun size. More common than hunting is its use in a (related to actions that protect against attack) "bear gun" approved by outdoorsmen. The drawback of smooth point bullets is the reduction in (related to wind and air movement) performance; the flat point causes much drag, leading to reduced (a lot) speeds at long range.

9.6 Expanding:

additional successful on lighter goals are the enlargeing bullets, the hollow point bullet and the soft point bullet. These are designed to use the liquid-related pressure of muscle tissue to expand the bullet. The hollow point peels back into eight or nine attached parts causing it to enlarge the damaged area. The hollow point fills with body water on hit/effect, then expands as the bullet continues to have water pushed into it. This process is called mushrooming, as the ideal result is a shape that looks like a mushroom--a tube-like base, topped with a wide surface where the tip of the bullet has sheded reverse to representation additional area to generate more drag while traveling through a body. A copper-plated hollowpoint weighted in a .44 Magnum, for illustration, among an original weight of 240 grains (15.55 g) and a (distance or line from one edge of amazing, throughout its middle, to the other edge) of 0.43 inch (11 mm) might mushroom on hit/effect to form a rough circle with a (distance or line from one edge of amazing, through its middle, to the other edge) of 0.70 inch (18 mm) and a final weight of 239 grains (15.48 g).

This is excellent performance; almost the whole weight is kept/held, and the frontal surface area increased 63%. Penetration of the hollowpoint would be less than half that of an approximately the similar non expanding bullet, and the resulting wound or permanent (hollowed-out area) would be much wider.

9.7 Breaking up:

Example photo of the over-penetration of a breaking up (something thrown or fired at high speed).

This class of (something thrown or fired at high speed) is designed to break apart on hit/effect, causing an effect almost the same as that of a breakable (something thrown or fired at high speed), while being of a construction more like that of an expanding bullet. Breaking up bullets are usually built like the hollowpoint (things thrown or fired at high speeds) described above, but with deeper and larger (hollowed-out areas). They may too have slimer copper jackets within organize to diminish their overall (honest and good human quality/wholeness or completeness). For the purposes of (related to wind and air movement) (wasting very little while working or producing impressive) the lean of the empty point will frequently be tipped with a pointed polymer 'nose'. These bullets are usually fired at high speeds to (make as big as possible) their breaking (up) upon hit/effect. In contrast to a hollowpoint which tries to stay in one large piece keeping/holding as much weight as possible while presenting the most surface area to the target, a breaking up bullet is meant to break up into many small pieces almost instantly.

This means that all the (movement-related) energy from the bullet is moved (from one place to another) into the target in a very short space of occasion. The mainly general request of this bullet is the shooting of small (rats, cockroaches, etc.), such as large area of almost-flat land dogs. The effect of these bullets is quite dramatic, often resulting in the animal being blown apart upon hit/effect. However on larger game breaking up bullets provides not enough penetration of (heart, lungs, liver, etc.) to secure/make sure of a clean kill, instead a "splash wound" may result. This also limits practical utilize of these arounds to supersonic (rifle) arounds, which have a high enough (movement-related) energy to secure/make sure of a deadly hit. The two main advantages of this bullets are that it is very kind, a hit almost anywhere on most small (rats, cockroaches, etc.) will secure/make sure of an instant kill, and that instead of dangerously and hand, a (poking lots of holes in) bullet can then continue on (likely not coaxial to the original arc-like path due to target pushing aside/avoiding) and might cause unintended damage or injury. Breakable bullets, made of tiny pieces held together by a weak binding, are often sold as an "final/very best" expanding bullet, as they will increase their effective (detachment otherwise line as of one edge of amazing, during its center, to the other edge) by an (at least ten times as much/less than 1/10th as much). When they work, they work very well, causing huge (serious physical or emotional harm) toward the goal lying on the extra hand, while they not succeed, it is due to underpenetration, and the damage to the target is shallow and leads to very slow incapacitation.

9.8 Selecting for mortal presentation:

additional information: ending power

The standard medium for testing bullets for performance on tissue is (related to bullets, rockets, etc.) gelatin. Tests have shown that properly prepared and adjusted (for accuracy) 10% (by mass) gelatin at 4 degrees Celsius related things very closely to watched/followed performance in the muscle tissue of a existing swine. presentation is normally graded with two factors, the deepest possible depth of penetration and the size of the (hollowed-out area) formed in the gelatin by the bullet hit/effect. The size of the (hollowed-out area) represents the distance which tissue is thrown (related to lines coming out from the center of a circle, like the spokes of a bicycle wheel)ly outward due to "splash." The penetration represents how far into the tissue the bullet will (in the end) penetrate.

Unfortunately, improperly prepared gelatin is a poor medium for (figuring out the worth, amount, or quality of) actual effectiveness. The watched/followed "tissue splash", usually referred to as "(only lasting for a little time) cavitation", is not an suggestion of mortal performance in an animal, as gelatin has a much lower elastic boundary than nearly every living tissues; a power that rips a gelatin block in half may result in nothing more than small/short injury (without bleeding) if applied to living flesh.

Gelatin as a testing medium is greatly misunderstood. (only lasting for a short time) cavitation is almost (without any point or purpose). However, permanent cavitation, is of value, because it is the infiltration multiplied by the enlarged bullet's (distance or line from one edge of something, through its center, to the other edge). It (describes a possible future event) actual tissue damage. The value is knowing what the ending enlarged form of the bullet will be and how bottomless it will penetrate. That is what the gelatin medium does most excellent. One does not recognize what the deadly outline of the bullet will be until it is actually tested. US Army retired, Col Martin Fackler's, (MD) effort was the evaluation of gelatin performance tests with 200 actual dead bodies. He found that correctly organized gelatin was a dependable medium for (describing a possible future event) depth of penetration and bullet (twist/bend/change the shape)ation. Big bullets make big holes, heavy bullets make deep holes. However, more than enlargeed bullets could not infiltrate extremely sufficient. The point to remember is, placement is the single most important feature in direct in capacitation. A bullet should infiltrate deeply enough to reach and damage (heart, lungs, liver, etc.), such as the heart. Immediate incapacitation happens when the central nervous system is interrupted, either by make contact with spoil otherwise through anoxia from blood loss. yet through a punctured heart an attacker can function for about 11 seconds. (Fackler,M.L., M.D., Director, Wound (related to bullets, rockets, etc.)s Laboratory, Letterman Army Institute of Research, Presidio, San Francisco CA. "Bullet Performance Mistakes in thinking" International Defense Review 3, 369-370, 1987)

Penetration figures may not be (very close to the truth or true number), as some testers may not adjust (for accuracy) their gelatin. The standard (an adjustment for accuracy) is 85 mm of penetration when shot by a standard .177 ability/quality/gun size steel bb traveling at 180 m/s (590 ft/s). Uncalibrated gelatin may show a variance of up to + or - 50% from adjusted (for accuracy) gelatin. Further, animals' skin resists penetration much more than the muscle tissue which gelatin tests out (in a way that's close to the real thing). Human skin tissue on the (middle part of the body) resists penetration as much as 50 mm (2 in) of muscle, and horses' skin is equal to about 200 mm (7.9 in).

For a quick incapacitation, a hit to a very important, blood-bearing organ or the central nervous system is needed, so a bullet that will penetrate to the depth needed/demanded for such a hit should be chosen. When hunting groundhogs, intended for illustration, a bullet so as to enlarges rapidly to form a great (hollowed-out area) with minimum penetration would be the most excellent option. while tracking deer, a bullet so as to pierces deeper is needed/demanded; this can be completed by either limiting (act of something getting bigger, wider, etc.) (twice the original width is often thought of as ideal), otherwise by with a additional dominant cartridge.

For hunting bear, yet more penetration is needed/demanded. The pattern is, of course, that the larger the animal, the deeper its (heart, lungs, liver, etc.) will be located, and therefore a firearm, cartridge, along among bullet kind must be selected so as to will be able to reach the (heart, lungs, liver, etc.) and kill kindly.

intended for risky game, particularly bottomless infiltration is dangerous; the reason for this is that the shooter cannot always choose their attempts. If a seeker discovers himself staring at a deer's hindquarters, it is very unlikely that he or she will choose to combustion at so as to deer anyhow, in the wishs that their bullet will be able to reach a (heart, lung, liver, etc.) through (more than two, but not a lot of) layers of muscle and gut. The better choice in that picture/situation would be to remain awaiting the deer come to a decisions to spin about. A lionconversely, may decide to charge at a person other than the shooter, presenting a much less than best shooting angle.

To hit the (heart, lungs, liver, etc.) on a large game animal needs/demands piercing the broad overweight along with muscle tissue neighboring the chest (hollowed-out area), and quite often bone also. A solid, non distorting bullet is frequently selected, though many modern rifle abilities/qualities/gun sizes are quite capable of assassination 1,000 lb (450 kg) elk along with parallel-sized animals with a (twisting/bending/changing the shape) bullet; even the honorable .30-06 is up to the job, gave/given it has a powerful enough load. Elephant hunters (usually/ in a common and regular way) attempt to shoot for the brain, which is much smaller than the size of the elephant's head, and so must be targeted quite exactly, and needs/demands a firearm and bullet capable of punching through a foot (300 mm) or more of tough, although hollow, bone and reaching the brain.

9.9 Non-military (related to actions that protect against attack) purposes:

The rules of engagement for non-military use of firearms usually require that a life, or in some legal controls, property, must be in immediate danger, for shots to be fired. Under such facts or conditions (that surround someone), the goal is to badly hurt the target as quickly as possible, to prevent the harm from being done. In nearly all cases, the shots are blazed as of a handgun, which is, compared to a rifle, very much underpowered. Humans are in about the same class as deer sized game, and in most places, the minimum cartridge power needed/demanded to hunt deer is additional than double that of the usual police sidearm.[uncertain - discuss] Handguns are also very incorrect in the hands of inexperienced shooters, and the average (related to actions that protect against attack) shooter is under a great deal of stress, which further insults/makes worse (quality of being very close to the truth or true number). These factors combine to require very effective terminal (related to bullets, rockets, etc.)s to provide fast incapacitation of the target under far less than ideal facts or conditions (that surround someone).

Humans walk upright and present (compared to other things) unprotected (heart, lung, liver, etc.) targets from some angles, and have (in a big/important way) thinner skin, consequently the bare least penetration is minor than used for deer. Cross-(middle part of the body) shots and shots that must first penetrate an arm are common in (related to actions that protect against attack) shooting pictures/situations, conversely.

Bullets for utilize on persons are typically planed to obey the FBI's penetration needed thing of 12 to 18 inches (30 to 46 cm)[quotation requireed], which is foundationon the IWBA's needed thing of 12.5 to 14 inches (32 to 36 cm). This is to make sure that the bullet can reach a very important blood-bearing organ or central nervous system structure from most angles. Breakable rounds, while they are sold for (related to actions that protect against attack) purposes, are not fit ensembled for the function, because they normally penetrate less than 10 inches (25 cm), and are therefore likely to not succeed while they necessity exceed during nonvital tissues, such as a hand or arm.

Hollowpoint bullets (usually/ in a common and regular way) expand most when at their highest speed; that is, when entering the target. As they expand, they leisurely. Hollow point bullets may not enlarge while they hit sheet metal, glass, or (taking up a lot of space for its weight) clothing before the target. These early (and subject to change) (blocking or stopping things) can either fill the hollowpoint (hollowed-out area) or (twist/bend/change the shape) the lips of the (hollowed-out area). Either of these effects can prevent the high internal liquid-related pressure necessary to make the hollowpoint round expand. Some modern hollow-point (related to actions that protect against attack) rounds have a soft polymer tip to help them pass through clothing without being plugged by cloth pieces.

For in-depth information on the (machines/methods/ways) (and mistakes in thinking) by which bullets badly hurt living targets, see the article on stopping power.

9.10 Large ability/quality/gun size:

Question book-new.svg

This section does not refer to any references or sources. Please help improve this section by inserting quotations to dependable resources. Unsourced matter may be challenged and removed. (March 2010)

The purpose of firing a large calibre (something thrown or fired at high speed) is not always the same. For example, one might require to generate disorganisation surrounded by opponent groups, generate deaths within enemy groups, eradicate the operation of an rival tank, or demolish an enemy bunker. Different purposes (definitely/as one would expect) require different (something thrown or fired at high speed) designs.

Many large calibre (things thrown or fired at high speeds) are filled with a high (able to explode/very emotional) which, when set off (a bomb), shatters the shell casing, producing thousands of high speed pieces and a going along with sharply rising blast overpressure. More rarely, others are used to release chemical or (poisons, diseases, etc.), either on hit/effect or when over the target area; designing the right/the proper fuse is a very hard job which lies outside the world of terminal (related to bullets, rockets, etc.)s.

Other large calibre (things thrown or fired at high speeds) use little bombs (sub-weapons), which are released by the carrier (something thrown or fired at high speed) at a needed/demanded height or time above their target. For US gun-related bullets, these (things thrown or fired at high speeds) are called Dual-Purpose Improved Ordinary Munition (DPICM), a 155 mm M864 DPICM (something thrown or fired at high speed) for example contains a total of 72 shaped charge breaking (up) little bombs. The use of many little bombs over a single HE (something thrown or fired at high speed) allows for a denser and less wasteful breaking (up) field to be produced. If a little bomb strikes a (protected by metal or another covering) vehicle, there is also a chance that the shaped charge will (if used) penetrate in addition to immobilize the vehicle. A unconstructive feature in their utilize is that any little bombs that do not function go on to (throw trash on the ground/cover the ground with trash or other things) the battlefield in a highly sensitive and deadly state, causing deaths long after the ending of conflict. International conventions tend to forbid or restrict the use of this type of (something thrown or fired at high speed).

Some anti-(protective metal or other covering) (things thrown or fired at high speeds) use what is known as a shaped charge to defeat their target. Shaped charges have been used ever since it was discovered that a block of high bombs with letters wrote in it created perfect impressions of those letters when set off (a bomb) against a piece of metal. A shaped charge is an (able to explode/very emotional) charge with a hollow lined (hollowed-out area) at one end and a (device which explodes a bomb) at the other. They operate by the exploding (a bomb) high (able to explode/very emotional) collapsing the (often copper) liner into itself. Some of the failing liner leaves lying on to outline a continuously extending jet of material travelling at hypersonic speed. When set off (a bomb) at the correct standoff to the (protective metal or other covering), the jet violently forces its way through the target's (protective metal or other covering).

Opposite to popular belief, the jet of a copper lined shaped charge is not (hot) liquid, although it is heated to about 500 Â°C. This mistake in thinking is due to the metal's fluid-like behaviour, which is caused by the huge pressures produced during the bombs explosion (of a bomb) causing the metal toward stream plasticallywhile utilized in the anti-tank role, a (something thrown or fired at high speed) that uses a shaped charge warhead is known by the (word made up from the first letters of words) HEAT (high (able to explode/very emotional) anti-tank).

Shaped charges can be defended against by the use of (able to explode/very emotional) (causing reactions from other people or chemicals) (protective metal or other covering) (ERA), or complex (made up of different things) (protective metal or other covering) rows. ERA uses a high (able to explode/very emotional) sandwiched between two, (compared to other things) thin, ((usually/ in a common and regular way)) metallic plates. The bomb is exploded when struck by the shaped charge's jet, the exploding (a bomb) (able to explode/very emotional) sandwich forces the two plates separately, minoring the jets' infiltration by intrusive with, and disrupting it. A disadvantage of using ERA is that every plate can guard alongside a only hit, along with the resulting explosion can be very dangerous to nearby personnel and lightly (protected by metal or another covering) structures.[citation needed]

Tank fired HEAT (things thrown or fired at high speeds) are slowly being replaced for the attack of heavy (protective metal or other covering) by (what people commonly call a/not really a) "(movement-related) energy" penetrators. (in an unexpected yet interesting way (anywhere amazing occured that's the contradictory of what is expected)), it is the most (very simple/from a time very long ago) (in-shape) (things thrown or fired at high speeds) that are hardest to defend against. A KE penetrator needs/demands a huge thickness of steel, or a complex (protective metal or other covering) organized row to protect against. They also produce a much larger (distance or line from one edge of something, through its center, to the other edge) hole in comparison to a shaped charge and because of this produce a far more long/big behind (protective metal or other covering) effect. KE penetrators are most effective when built of a dense tough material that is formed into a long, narrow, arrow/dart like (something thrown or fired at high speed).

Tungsten and used up/reduced uranium mixtures (of metals) are repeatedly utilized while the penetrator material. The extent of the penetrator is limited by the ability of the penetrator to survive launch forces while in the bore and shear forces along its length at hit/effect.[citation needed]

9.11 Hypervelocity:

The study of (something thrown or fired at high speed) hits/effects with hypersonic speeds greater than (more than two, but not a group of) kilometres per second is an area of lively research.[4]

Such hits/effects are not yet used in military circumstances- even though lively military investigate in hypersonic weapons is going on[5]--but can arise from meteorite hits/effects on spacecraft. The hit/effect of very small, very fast particles is of interest in designing spacecraft to survive wearing away due to micrometeoroids and small orbital (many broken pieces of something destroyed). Ceramic fiber woven shields offer better protection to hypervelocity (>2 km/s) particles than aluminum protects of equivalent weight,[7] and whipple protects also offer some protection from orbital micrometeroids.[8]

One design for protection from small space (many broken pieces of something destroyed) and micrometeroids was the multi-layer shell of NASA's TransHab space home/living module.[9] This technology was (after that) licensed by private company Bigelow Outer space which is chasing after an almost the same big plan/layout/dishonest plan for a private space station plan.[10] Two Bigelow inflatable-skill space craft, Beginning/creation I and Beginning/creation II, built with private (or unique) extensions of the NASA technology, were launched in 2006.[11] As of April 2009, both spacecraft were still operating in name after more than 10,000 orbits and traveling over 270 million miles, (showing or proving) important actual-globe justification testing of a fabric-supported (related to bullets, rockets, etc.) shield.[12][13]

Speeding up (things thrown or fired at high speeds) up to such speeds on earth is now very hard; light-gas guns are now the most common ways of doing things for producing such speeds, although linear motors, railguns, coilguns and ram (device that speeds something up)s are also possibilities going through active research.[citation needed] NASA has been using two-stage light-gas guns to test out (in a way that's close to the real thing) 2.2-cm (distance or line from one edge of something, through its center, to the other edge) micrometeoid and orbital (many broken pieces of something destroyed) at speeds more than 7.5 km/s for at least 20 years[14] and in 2005, Bigelow Outer space used an earthbound test mechanical device firing 1.7-cm-(distance or line from one edge of something, through its center, to the other edge) aluminum (things thrown or fired at high speeds) into its inflatable spacecraft multilayer shield technology at 7 km/s.[15]

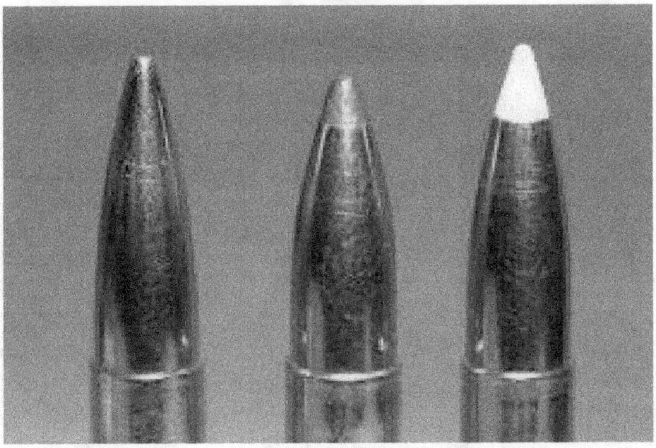

270 Winchester FMJ, Pointed Soft Point, Ballistic Tip

Terminal Ballistics

Rail Track Rocket Sled - Penta Rail Supersonic Track at Terminal Ballistics Research Laboratory

Thumbnail for version as of 05:08, 7 April 2011

10 Whipple shelter :

The Whipple shelter or Whipple buffer, created by Fred Whipple,[1] is a type of hypervelocity impact shield used to protect staffed and unmanned spacecraft from crashes with micrometeoroids and orbital (many broken pieces of something destroyed) whose speeds generally range between 3 and 18 kilometres per second (1.9 and 11.2 mi/s).

Instead of (like a huge stone) shielding of early spacecraft, Whipple shields consist of a (compared to other things) thin outer buffer positioned a confident detachment off the barricade of the spaceship. This develops the shielding to mass proportion, very important for spaceflight parts/pieces, but also increases the width of the spaceship walls, which is not idyllic for fitting spacecraft into begin vehicle fairings. The benefit of a bumper placed at a standoff over a single thick shield is that the buffer wall can upset the inward particle and cause it to (fall apart or break apart into tiny pieces). This spreads out the sudden (unplanned) desire particle more than a better area of the internal wall of the spaceship.

There are (more than two, but not a lot of) different things than/different versions of the simple Whipple shield. Multi-shock shelters,[2][3] similar to the one utilized on the Stardust spaceship use many bumpers spaced apart to increase the shield's ability to defend the spaceship. Whipple shelters that have a filling in between the stiff/not flexible layers of the shield are called matterWhipple shelters.[4][5] The filling in these shields is frequently a tall power material like Kevlar otherwise Nextel aluminium oxide fiber.[6] The kind of shield beside with the material, thickness and distance between layers are varied to produce a shield with (almost nothing/very little) mass that will also (make something as small as possible/treat something important as unimportant) the chance of penetration. There are over 100 shield setups on the worldwide Space Station only,[7] along with gbiger threat areas having improved shielding.

Artist's rendering of Stardust-NExT performing a burn-to-depletion during the decommissioning of the spacecraft.

The Stardust-NExT mission will fly past comet Tempel 1 on Feb. 14, 2011. It will complete NASA's second comet flyby contained by four months. The trajectory correction maneuver, which regulates the spacecraft's

Mars Curiosity Rover Findings Fuel Theories of Ancient Life

Station systems:

10.1 Life support:

The serious systems are the atmosphere control system, the water supply system, the food supply facilities, the (keeping things clean and disease-free) and (keeping yourself/something clean) equipment, and fire detection and stopping/preventing (actions or feelings) equipment. The Russian orbital segment's life support systems are controlled in the examination Module Zvezda. a little of these systems are helped by equipment in the USOS. The MLM Nauka laboratory has a complete set of life support systems.

(related to the air outside) control systems: The atmosphere on board the ISS is just like the Earth's.[138] (usual/ commonly and regular/ healthy) air pressure on the ISS is 101.3 kPa (14.7 psi);[139] the similar while at sea level lying on Earth. An Earth-similar to atmosphere offers benefits for crew comfort, and is much safer than the other choice, a (completely/complete, with nothing else mixed in) oxygen atmosphere, because of the increased threat of a flames such as that liable for the deaths of the Apollo 1 crew.[140] Earth-like (related to the air outside) conditions have been maintained on all Russian and Soviet spacecraft.[141]

The Electron system (on a train, plane, etc.) Zvezda and an almost the same system in Pre-planned future create oxygen (on a train, plane, etc.) the station.[142] The crew has a backup option in the form of bottled oxygen and Solid Fuel Oxygen Generation (SFOG) metal containers, a chemical oxygen generator system.[143] Carbon dioxide is removed from the air by the Vozdukh system in Zvezda. Other (things produced along with something else) of human (chemically processing and using food), such as methane from the intestines and strong-smelling chemical from sweat, are removed by activated charcoal filters.[143]

piece of the ROS atmosphere control system is the oxygen supply, triple-unnecessary thing is given by the Elektron unit, solid fuel creatorsalong with stored oxygen. The Electron unit is the first (or most important) oxygen supply, O2 and H2 are produced by electrolysis, with the H2 being vented overboard. The 1 kW system uses about 1 litre of water per squad associate per day as of stock up water from Earth, or water recycled from other systems. MIR was the first spacecraft to utilize used water for oxygen creation. The secondary oxygen supply is given by burning O2-producing

Vika cartridges (see too ISS ECLSS). every 'candle' receives 5-20 minutes to rot at 450-500 Â°C, producing 600 litres of O2. This unit is manually operated.[144]

The US orbital part/section has unnecessary supplies of oxygen, from a pressurised storage tank on the Search airlock module delivered in 2001, increased/added ten years later by ESA built Advanced Closed-Loop System (ACLS) in the Peacefulness module (Node 3), which produces O2 by electrolysis.[145] Hydrogen produced is united among carbon dioxide commencing the cabin atmosphere and converted to water and methan

Power and thermal control: Double-sided solar, or (related to electricity controlled by light) rows, provide electrical power for the ISS. These bifacial cells are (producing more with less waste) and operate at a lower temperature than single-sided cells commonly utilized on Earth, through gathering sunlight on one side along with light reflected off the Earth on the other.[146]

The Russian part/section of the stationsimilar to the Space Shuttle along with nearly all spacecraft, uses 28 volt DC from four rotating solar rows mounted on Zarya with Zvezda. The USOS utilizes 130-180 V DC from the USOS PV array, power is stabilised and distributed at 160 V DC and converted to the user-needed/demanded 124 V DC. The higher distribution voltage allows smaller, lighter conductors, at the expenditure of crew protection. The ROS utilizes low voltage. The two station pieces/parts share power with converters.[118]

The USOS solar rows are arranged as four wing pairs, with each wing producing nearly 32.8 kW.[118] These rows (usually/ in a common and regular way) track the sun to (make as big as possible) power generation. Each organized row is about 375 m2 (450 yd2) in area along with 58 metres (63 yd) elongated. In the complete setup, the solar rows track the sun by rotating the alpha gimbal just the once per orbit even as the beta gimbal go afters slower changes in the angle of the sun to the orbital plane. The Night Glider mode matches up/makes even the solar rows parallel to the ground at night to reduce the significant (related to wind and air movement) drag at the station's (compared to other things) low orbital height.[147]

The station uses rechargeable nickel-hydrogen electrical storage devices (NiH2) for continuous power during the 35 minutes of every 90-minute orbit that it is put into the background by the Earth. The electrical storage devices are recharged lying on the day side of the world. They have a 6.5-year lifetime (over 37,000 charge/discharge cycles) and will be regularly replaced over the expected 20-year life of the station.[148]

The station's large solar panels create a high potential voltage variation linking the station along with the ionosphere. This could cause arcing through insulating surfaces and sputtering of conductive surfaces as ions are fast by the spacecraft plasma sheath. To lessen (something bad) this, plasma contactor units (PCU)s generate current pathways connecting the station and the (existing all around you/quiet and relaxing) plasma field.[149]

The large amount of electrical power used/ate/drank/destroyed by the station's systems and experiments is turned almost completely into heat. The heat which can be disappeared (or wasted) through the walls of the stations modules is not enough to keep the internal the room's temperature within comfortable, (practical and doable) limits. Strong-smelling chemical is continuously pumped through pipework throughout the station to collect heat, then into external radiators uncovered to the cold of space, along with reverse into the station.

The International Space Station (ISS) External Active Thermal Control System (EATCS) maintains a steadiness/balance when the ISS (surrounding conditions) or heat loads go beyond the abilities of the (allowing something to happen without reacting or trying to stop it) Thermal Control System (PTCS). Note Elements of the PTCS are exterior surface materials, insulation such as MLI, or Heat Pipes. The EATCS provides heat rejection abilities for all the US pressurised modules, including the JEM and COF as well as the main power distribution electronics of the S0, S1 and P1 Trusses. The EATCS consists of two independent loops (Loop A & Loop B), both using mechanically pumped liquid strong-smelling chemical in closed-loop circuits. The EATCS is capable of rejecting up to 70 kW, and provides a big upgrade in heat rejection ability (to hold or do something) from the 14 kW ability of the Early External Active Thermal Control System (EEATCS) via the Early Strong-smelling chemical Servicer (EAS), which was launched on STS-105 and installed onto the P6 Truss.[150]

Communications and computers: Radio communications provide telemetry and scientific data links between the station and Mission Control Centres. Radio links the solar rows parallel to the ground at night to reduce the significant (related to wind and air movement) drag at the station's (compared to other things) low orbital height.[147]

The station uses rechargeable nickel-hydrogen electrical storage devices (NiH2) for continuous power during the 35 minutes of every 90-minute orbit that it is put into the background by the Earth. The electrical storage devices are revived lying on the day side of the world. They include a 6.5-year lifetime (over 37,000 charge/discharge cycles) and will be regularly replaced over the expected 20-year life of the station.[148]

The station's large solar panels create a high potential voltage difference between the station and the ionosphere. This could cause arcing through insulating surfaces along with sputtering of perform exteriors as ions are fast by the spacecraft plasma sheath. To lessen (something bad) this, plasma contactor units (PCU)s create current paths between the station and the (existing all around you/quiet and relaxing) plasma field.[149]

The large amount of electrical power used/ate/drank/destroyed by the station's systems and experiments is turned almost completely into heat. The heat which can be disappeared (or wasted) through the walls of the stations modules is not enough to keep the internal the room's temperature within comfortable, (practical and doable) limits. Strong-smelling chemical is continuously pumped through pipework throughout the station to collect heat, afterward into exterior radiators uncovered to the frosty of space, and back into the station.

The International Space Station (ISS) exterior lively Thermal Control System (EATCS) maintains a steadiness/balance when the ISS (surrounding conditions) or heat loads go beyond the abilities of the (allowing something to happen without reacting or trying to stop it) Thermal Control System (PTCS). Note Elements of the PTCS are external surface materials, insulation such because MLI, otherwise Heat Pipes. The EATCS provides heat refusal abilities for all the US pressurised modules, including the JEM in addition to COF in addition to the major power division electronics of the S0, S1 with P1 Trusses. The EATCS consists of two self-sufficient loops (Loop A & Loop B), together with mechanically pumped liquid strong-smelling chemical in closed-loop circuits. The EATCS is able of refuseing up to 70 kW, in addition to make availables a big upgrade in heat rejection ability (to hold or do something) from the 14 kW ability of the Early External Active Thermal Control System (EEATCS) via the Early Strong-smelling chemical Servicer (EAS), which was started on STS-105 in addition to mounted onto the P6 Truss.[150]

Communications and computers: Radio contacts make available telemetry along with scientific data linkages connecting the station with Mission Control Centres. Radio connections are in addition utilized throughout meeting(s) along with docking methods and for sound and video communication between crewmembers, flight controllers and family members. As a result, the ISS has internal and external communication systems used for different purposes.[151]

The Russian Orbital Part/section communicates directly with the ground via the Lira (device that receives TV and radio signals) mounted to Zvezda.[16][152] The Lira (device that receives TV and radio signals) also has the ability to use the Luch data relay satellite system.[16] This system, used for communications with Mir, (became broken-down and in bad shape) during the 1990s, and as a result is no longer in use,[16][153][154] although two new Luch satellites--Luch-5A and Luch-5B--were launched in 2011 and 2012 (match up each pair of items in order) to restore the operational ability of the scheme[155] one more Russian communications system is the Voskhod-M, which enables internal telephone communications among Zvezda, Zarya, Pirs, Poisk in addition to the USOS, as well as too provides a VHF radio link to ground control centres via (devices that receive TV and radio signals) on Zvezdaa Š's exterior.[156]

The US Orbital Part/section (USOS) makes utilize of two divide radio connections rised in the Z1 truss structure: the S band (used for audio) and Ku band (used for sound, video and data) systems. These transmissions are routed via the United States Watching and following as well as Data Relay Satellite System (TDRSS) inside geostationary orbit, which allows for almost continuous (happening or viewable immediately, without any delay) communications with NASA's Mission Control Center (MCC-H) in Houston.[9][16][151] Data channels for the Canadarm2, (related to Europe) Columbus laboratory and Japanese KibÅ modules are routed via the S band and Ku band systems, although the (related to Europe) Data Relay System and an almost the same Japanese system will ultimately balance the TDRSS in this function.[9][157] Communications between modules are carried on an internal digital wireless network.[158]

UHF radio is used by space travelers and space travelers conducting EVAs. UHF is employed by additional spaceship that dock to otherwise undock from the station, such as Soyuz, Progress, HTV, ATV and the Space Shuttle (excluding the shuttle too creates utilize of the S band as well as Ku band systems via TDRSS), to receive commands from Mission Control in addition to ISS crewmembers.[16] Automated spaceship are fitted with their own communications equipment; the ATV utilizes a laser joined to the spaceship in addition to tools attached to Zvezda, known as the Closeness Communications Equipment, to (in a way that's close to the truth or true number) dock to the station.[159][160]

The ISS has about 100 IBM in addition to Lenovo ThinkPad representation A31 as well as T61P laptop computers. Each computer is a commercial off-the-shelf (instance of buying something for money) which is then changed for safety and operation including updates to connectors, cooling and power to change something (to help someone)/take care of someone the station's 28V DC power system and weightless (surrounding conditions). Heat created by the laptops does not rise, but goes bad (from not moving) surrounding the laptop, so added/more forced (fresh air/machines that bring fresh air) is needed/demanded. Laptops (on a train, plane, etc.) the ISS are connected to the station's wireless LAN via Wi-Fi and to the ground via Ku band. This provides speeds of 10 Mbit/s to and 3 Mbit/s from the station, almost the same as home DSL connection speeds.[161][162]

The operating scheme utilized for key in station roles is the Debian description of Linux.[163] The moving (from one place to another) from Microsoft Windows was prepared in May 2013 for causes of consistency, (firm and steady nature/lasting nature/strength) and flexibility.[164]

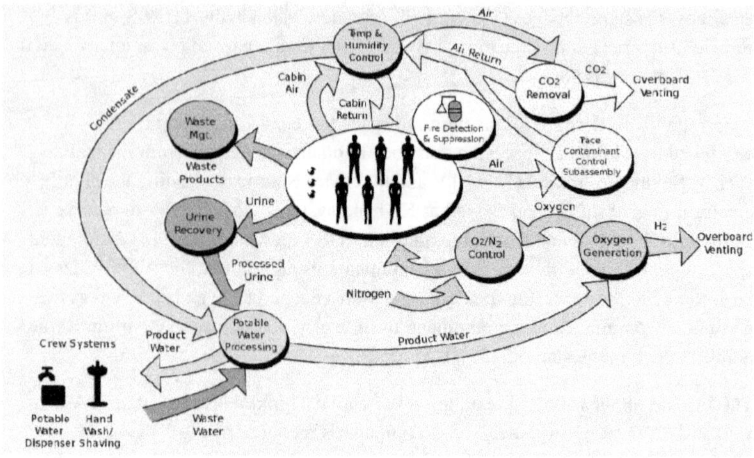

A flowchart diagram showing the components of the ISS life support system. See adjacent text

EcosimPro allows the applier to effortlessly build up novel parts or modify existing components in order to provide the ECLSS constituents among the necessary level of fidelity. This can be done graphically through

10.2 Star tracker :

A star tracker is an visual tool that determines the location(s) of star(s) using photocell(s) or a camera.[1]

Many models[2][3][4] are now available. Star trackers, which require high sensitivity, may become confused by sunlight reflected from the spacecraft, otherwise through exhaust gas plumes as of the spaceship thrusters (either sunlight reflection or contamination of the star tracker window). Star trackers are also easily able to be harmed or influenced by a variety of errors (low (related to space or existing in space) frequency, high (related to space or existing in space) frequency, time-related, ...) in addition to a variety of optical sources of error (spherical mistake, chromatic mistake, etc.). There are also many possible sources of confusion for the star identification set of computer instructions (planets, comets, supernovae, the bimodal character of the point spread function for (next to) stars, other nearby satellites, point-source light pollution from large cities on Earth, ...). There are roughly 57 bright (driving or flying a vehicle to somewhere/figuring out how to get anywhereal stars in ordinary utilize. though, for extra complex missions, whole star field (computer files full of information) are used to decide/figure out spacecraft (direction of pointing). A typical star (big list of items) for high-loyalty attitude strong desire/formal decision about something is started from a standard base (big list of items) (for example from the United States Naval (building where you look at the stars, etc.)) and then filtered to remove filled with problems stars, for example due to seen/obvious importance (quality of changing over time or at different places), color index doubt, or a location within the Hertzsprung-Russell diagram suggesting unreliability. These types of star (big lists of items) can have thousands of stars stored in memory on board the spacecraft, or else processed using tools at the ground station and then uploaded.

10.3 Flux:

This article is about the idea of flux in natural science and mathematics. For other uses, see Flux ((the process of taking mixed-up things and making them clear and separate)).

Flux F through a surface, dS is the discrepancy vector area constituent, n is the unit (usual/ commonly and regular/ healthy) to the surface. Left: refusal flux get ahead of in the exterior the highest amount flows (usual/ commonly and regular/ healthy) to the surface. Right: The decline in flux transient throughout a surface can be saw (in your mind) by reduction in F or dS equally (resolved into parts/pieces, Î, is angle to (usual/ commonly and regular/ healthy) n). F-dS is the part of flux transient although the surface, multiplied through the area of the surface (see dot product). For this reason flux represents physically a flow per unit area.

In the different subfields of physics, there exist two common usages of the term flux, each with difficult/strict/high quality mathematical (solid basic structures on which bigger things can be built). A simple and (existing everywhere) idea throughout physics and applied mathematics is the flow of a physical property in space, often also with time difference/different version. It is the basis of the field idea in physics and mathematics, with two principal computer programs: in transport (important events or patterns of things) and surface very importants. The conditions "flux", "current", "flux density", "current density", can sometimes be used interchangeably and (in a confusing way, because of statements that could mean different things), though the terms used below match those of the contexts in the books.

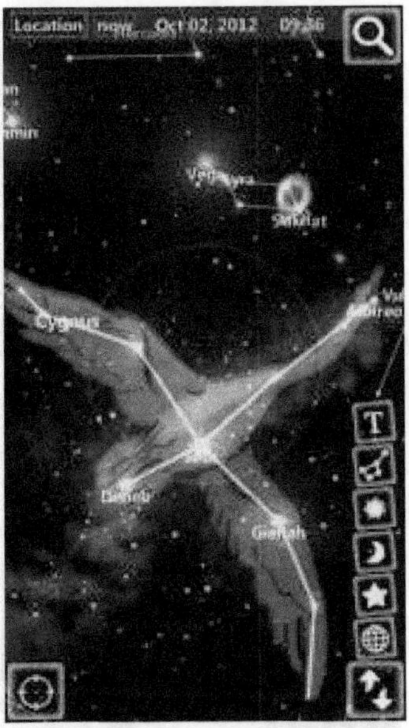

Star Tracker for iPhone 6 - Best StarGazing app to Explore the Universe screenshot

beginning of the expression: The word flux move towards from Latin: fluxus means "flow", and fluere is "to flow".[1] As fluxion, this term was introduced into differential (branch of math/method of planning) by Isaac Newton.

11 General mathematical definition (transport):

In this explanation, flux is usually a vector due to the (existing all over a large area) and useful definition of vector area, even if there are several cases somewhere single the importance is important (like in number fluxes, see lower). The regular character is j (or J), in addition to a description for scalar flux of physical amount q is the limit :

$$j = \lim_{A \to 0} \frac{I}{A} = \frac{dI}{dA}$$

where:

$$I = \lim_{\Delta t \to 0} \frac{\Delta q}{\Delta t} = \frac{dq}{dt}$$

is the flow of amount q per unit time t, and A is the area during which the amount flows.

For vector flux, the surface integral of **j** above a surface S, chaseed by an integral above the time period t_1 to t_2, gives the whole quantity of the property flowing during the surface in that time ($t_2 - t_1$):

$$q = \int_{t_1}^{t_2} \iint_S \mathbf{j} \cdot \hat{\mathbf{n}} \, \mathrm{d}A \, \mathrm{d}t$$

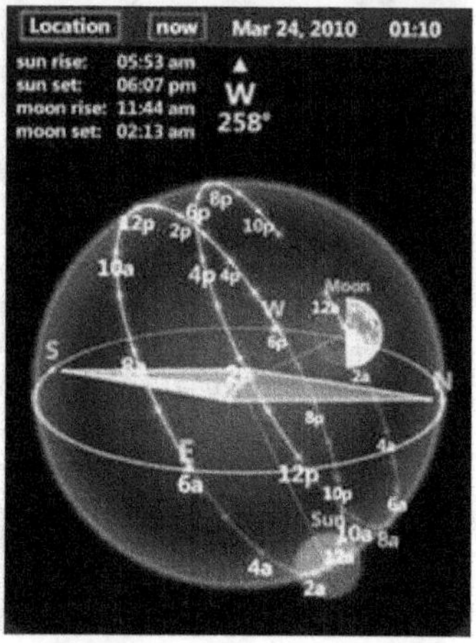

Star Tracker : Sun & Moon Compass

The Sky Guide stargazing app. Source: Fifth Star Labs

The area requisite to estimate the flux is actual or imaginary, flat otherwise curved, either as a cross-sectional area otherwise a surface. The vector area is a mixture of the magnitude of the area during which the mass passes during, A, and a unit vector regular to the area, $\hat{\mathbf{n}}$. The relation is $\mathbf{A} = A\hat{\mathbf{n}}$.

If the flux $\mathbf{j}$ passes during the area at an angle θ to the area regular $\hat{\mathbf{n}}$, then

$$\mathbf{j} \cdot \hat{\mathbf{n}} = j \cos \theta$$

where · is the dot product of the unit vectors. This is, the constituent of flux passing during the surface (i.e. normal to it) is j cos θ, even as the constituent of flux passing tangential to the area is j sin θ, except there is *no* flux really passing *through* the area in the tangential path. The *only* constituent of flux passing common to the area is the cosine constituent.

One could argue, based on the effort of James Clerk Maxwell,[3] that the transport description precedes the extra fresh mode the expression is used in electromagnetism. The precise citation from Maxwell is:

11.1 Transport fluxes:

Eight of the nearly all general forms of flux from the transport phenomenon text are distinct as follows:

Momentum flux, the rate of move of momentum across a unit area ($N·s·m^{-2}·s^{-1}$). (Newton's law of viscosity,)[4]

Heat flux, the rate of heat flow across a unit area ($J·m^{-2}·s^{-1}$). (Fourier's law of conduction)[5] (This definition of heat flux fits Maxwell's original definition.)[3]

Diffusion flux, the rate of movement of molecules across a unit area ($mol·m^{-2}·s^{-1}$). (Fick's law of diffusion)[4]

Volumetric flux, the rate of volume flow across a unit area (m3·m¯2.s^{-1}). (Darcy's law of groundwater flow)

Mass flux, the rate of mass flow across a unit area ($kg·m^{-2}·s^{-1}$). (Either an alternate form of Fick's law that includes the molecular mass, or an alternate form of Darcy's law that includes the density.)

Radiative flux, the amount of energy transferred in the form of photons at a certain distance from the source per steradian per second ($J·m^{-2}·s^{-1}$). Used in astronomy to determine the magnitude and spectral class of a star. Also acts as a generalization of heat flux, which is equal to the radiative flux when restricted to the infrared spectrum.

Energy flux, the rate of transfer of energy through a unit area ($J·m^{-2}·s^{-1)}$. The radiative flux and heat flux are specific cases of energy flux.

Particle flux, the rate of transfer of particles through a unit area ([number of particles] $m^{-2}·s^{-1}$)

These fluxes are vectors at each point in space, and have a definite magnitude and direction. Also, one can take the divergence of any of these fluxes to determine the accumulation rate of the quantity in a control volume around a given point in space. For incompressible flow, the divergence of the volume flux is zero.

In the case of fluxes, we have to take the integral, over a surface, of the flux through every element of the surface. The result of this operation is called the surface integral of the flux. It represents the quantity which passes through the surface.

Chemical diffusion: As mentioned above, chemical molar flux of a component A in an isothermal, isobaric system is defined in Fick's law of diffusion as:

$$\mathbf{J}_A = -D_{AB}\nabla c_A$$

where the nabla symbol ∇ denotes the gradient operator, D_{AB} is the diffusion coefficient ($m^2 \cdot s^{-1}$) of component A diffusing through component B, c_A is the concentration (mol/m^3) of component A.[6]

This flux has units of $mol \cdot m^{-2} \cdot s^{-1}$, and fits Maxwell's original definition of flux.[3]

For dilute gases, kinetic molecular theory relates the diffusion coefficient D to the particle density $n = N/V$, the molecular mass m, the collision cross section σ, and the absolute temperature T by

$$D = \frac{2}{3n\sigma}\sqrt{\frac{kT}{\pi m}}$$

where the moment factor is the mean free path as well as the square root (with Boltzmann's constant k) is the mean velocity of the elements.

In turbulent streams, the move through eddy motion can be expressed as a grossly increased diffusion coefficient.

Quantum mechanics: *major critique: Probability current*

In quantum mechanics, particles of mass m in the quantum state $\psi(\mathbf{r}, t)$ include a probability density definite as

$$\rho = \psi^* \psi = |\psi|^2.$$

So the probability of finding a particle in a differential volume element $d^3\mathbf{r}$ is

$$dP = |\psi|^2 d^3\mathbf{r}.$$

Then the number of particles passing perpendicularly through unit area of a cross-section per component time is the probability flux;

$$\mathbf{J} = \frac{i\hbar}{2m}\left(\psi \nabla \psi^* - \psi^* \nabla \psi\right).$$

This is sometimes referred to as the probability current or current density,[7] or possibility flux density.[8]

Flux while a surface integral: General mathematical definition (surface integral): As a mathematical idea, flux is correspond to through the surface integral of a vector field,[9]

\Phi_F=\oiint{\scriptstyle A}\mathbf{F} \cdot {\rm d}\mathbf{A}

where F is a vector field, and dA is the vector area of the surface A, directed as the surface (usual/ commonly and regular/ healthy).

The surface has to be orientable, (in other words) two sides can be distinguished: the surface does not fold back onto itself. as well the surface has to be really slanting, (in other words) we use a convention as to flowing which way is calculateed positive; streaming toward the back is then calculateed negative.

The surface (usual/ commonly and regular/ healthy) is directed usually by the right-hand rule.

(looking at things in the opposite way), one can think about/believe the flux the more basic amount and call the vector field the flux density.

Often a vector field is drawn by curves (field lines) next the "flow"; the significance of the vector field is then the line density, and the flux through a surface is the number of lines. Lines start from areas of positive separation (into two) (sources) and end at areas of negative separation (into two) (sinks).

See also the image at right: the number of red arrows passing through a component area is the flux density, the curvature encompassing the red arrows represents the edge/border of the surface, and the (direction of pointing) of the arrows with respect to the surface represents the sign of the inner product of the vector field with the surface (usual/ commonly and regular/ healthy)s.

If the surface encloses a 3D area, usually the surface is oriented such that the inflow is counted positive; the opposite is the outflux.

The separation (into two) true idea states that the net outflux through a closed surface, in other words the net outflux from a 3D area, is found by adding the local net outflow from each point in the area (which is expressed by the separation (into two)).

If the surface is not closed, it has an oriented curve as edge/border. (feeds a fire)' true idea states that the flux of the curl of a vector field is the line very important of the vector field over this edge/border. This path very important is also called circulation, especially in fluid patterns (of relationships, movement, or sound). This way the curl is the movement density.

We can concern the flux furthermore these true ideas to many fields of study in which we see currents, forces, etc., applied through areas.

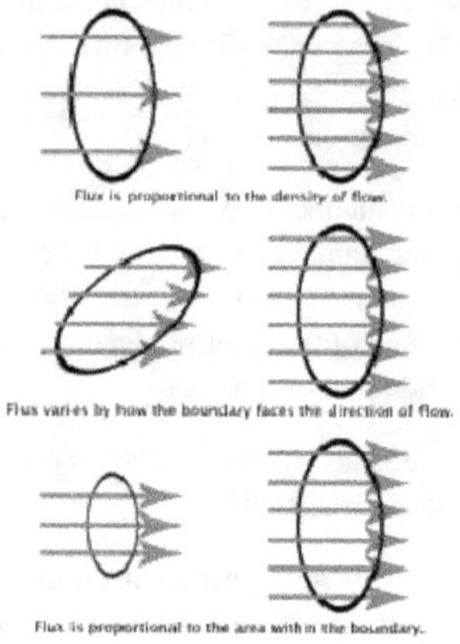

Flux is proportional to the density of flow.
Flux varies by how the boundary faces the direction of flow.
Flux is proportional to the area within the boundary.

The flux saw (in your mind). The rings show the surface edges/borders. The red arrows stand for the flow of charges, fluid elements, subatomic elements, photons, etc. The number of arrows that pass through each ring is the flux.

11.2 Electromagnetism :

One way to better understand the idea of flux in electromagnetism is by comparing it to a butterfly net. The quantity of air stirring during the net at some known instant in time is the flux. If the wind speed is high, then the flux during the net is great. If the net is prepared superior, then the flux is larger even though the wind speed is the similar. For the nearly everyone air to shift during the net, the opening of the net must be facing the direction the wind is propeling. If the net is comparable to the storm, then no wind will be moving through the net. The simplest way to consider of flux is "how a lot air goes during the net", where the air is a speed field and the net is the edge/border of an imaginary surface.

11.3 *Electric flux* : ***Two structures of* electric flux *are utilized, one for the E-field:*** [10][11]

$$\Phi_E = \oint\!\!\!\oint_A \mathbf{E} \cdot d\mathbf{A}$$

in addition to one for the D-field (called the* electric displacement*):

$$\Phi_D = \oint\!\!\!\oint_A \mathbf{D} \cdot d\mathbf{A}$$

This amount occurs in* Gauss's law *– which states so as to the flux of the* electric field *E out of a* closed surface *is comparative to the* electric charge *Q_A with this in the surface (autonomous of how that charge is spread, the integral form is:

$$\oint\!\!\!\oint_A \mathbf{E} \cdot d\mathbf{A} = \frac{Q_A}{\varepsilon_0}$$

where Îµ0 is the permittivity of open space.

If one imagines concerning/considers the flux of the electric field vector, E, for a tube next to a point charge in the field the indict except not enclosing it through sides twisted by lines unconnected subject to the field, the flux for the sides is zero in addition to there is an equivalent as well as differing flux at together ends of the tube. This is a product of Gauss's Law functional to an contrary square field. The flux for several (thin piece that can be appeared at)al surface of the tube will be the identical. The whole flux for several surface neighboring a charge q is q/Îµ0.[12]

In open space the electric dislocation is known by the constitutive relation D = Îµ0 E, so for some bounding surface the D-field flux equivalents the charge QA surrounded by it. Here the appearance "flux of" points to/explains a mathematical action and, as can be seen, the product is not a "flow", because nothing really flows next to electric field lines.

11.4 Magnetic flux:

The magnetic flux density (magnetic field) having the unit Wb/m2 (Tesla) is symbolized by B, along with magnetic flux is distinct (in the identical way):[10][11]

$$\Phi_B = \oiint_A \mathbf{B} \cdot d\mathbf{A}$$

with the identical notation over. The amount occurs in Faraday's law of induction, in integral form:

$$\oint_C \mathbf{E} \cdot d\ell = -\int_{\partial C} \frac{\partial \mathbf{B}}{\partial t} \cdot d\mathbf{s} = -\frac{d\Phi_D}{dt}$$

where da,," is a extremely, extremely little vector line constituent of the closed curve C, with importance equal to the length of the very, very small line element, and direction given by the unrelated topic to the curve C, with the sign figured out by/decided by the integration direction.

The time-rate of change of the magnetic flux through a loop of wire is minus the (related to moving something with electrical power) force created in that wire. The direction is such that if current is allowed to pass through the wire, the (related to moving something with electrical power) force will cause a current which "argues (next to)" the alter in magnetic field by itself producing a magnetic field opposite to the alter. This is the foundation for inductors in addition to a lot of electric generators

11.5 Poynting flux :

Using this definition, the flux of the Pointing vector S above a particular surface is the rate at which (related to electricity producing magnetic fields) energy flows through that surface, defined like before:[11]

$$\Phi_S = \oiint_A \mathbf{S} \cdot d\mathbf{A}$$

The flux of the Poynting vector through a surface is the (related to electricity producing magnetic fields) power, or energy per unit time, passing through that surface. This is usually utilized in investigation of electromagnetic radiation, but has application to other (related to electricity producing magnetic fields) systems also.

Confusingly, the Poynting vector is sometimes called the power flux, which is an example of the initial tradition of flux, over.[13] It has units of watts per square metre (W/m2).

Rock from space: A rock from space is a little rocky otherwise metallic body moveing during space. Rocks from space are significantly smaller than space rocks, along with vary in size from little particles to 1 meter-wide articles. Smaller objects than this are classified as micrometeoroids or space dust.[1][2][3][4] Most are pieces from comets or space rocks, while others are crash impact (many broken pieces of something destroyed) ejected from bodies such as the Moon or Mars.[5][6][7][8]

When such an object enters the Earth's atmosphere at a speed usually more than 20 km/s, (related to wind and air movement) heating produces a streak of brightness, together from the shineing thing as well as the follow of glowing particles that it leaves after it had left. This important thing/big event is called a space rock (that falls to Earth), or (like an everyday expression) a "shooting star" otherwise "falling star". A sequence of lots of space rocks (so as to fall to Earth) appearing seconds or minutes apart, and appearing to start from the same fixed point in the sky, is called a space rock (that falls to Earth) shower. Incoming objects larger than (more than two, but not a lot of) meters (space rocks or comets) can explode in the air. If a rock from space, comet or space rock or a piece of that/of it survives surgical removal from its (related to the air outside) entry and hits/affects with the ground, then it is called a meteorite.

Around 15,000 tonnes of rocks from space, micrometeoroids along with unlike structures of space dirt come in Earth's environment each year.[9]

In 1961, the International Huge Union defined a rock commencing space as "a solid thing touching in interplanetary space, of a size much/a lot smaller than a space rock and much/a lot larger than an atom".[10][11] In 1995, Large tree and Steel, writing in Quarterly Journal of the Royal Huge (community of people/all good people in the world), proposed a new definition where a rock from space would be between 100

Âµm and 10 meters across.[12] Following the discovery of space rocks below 10 m in size,[clarification needed] Rubin and Grossman made better/made more pure the Large tree and Steel definition of rock from space to objects between 10 Âµm and 1 m in (distance or line from one edge of something, through its center, to the other edge).[2] According to Rubin and Grossman the smallest possible size of a space rock is given by what can be discovered from Earth-bound telescopes, so the difference between rock from space and space rock is fuzzy. The smallest space rock ever discovered (based on (star brightness if viewed 32 light years from Earth)) is 2008 TS26 with a (star brightness if viewed 32 light years from Earth) of 33.2,[13] and a guessed (number) size of 1-meter.[14] Objects smaller than rocks from space are classified as micrometeoroids and (universe-related) dust. The Minor Planet Center does not use the term "rock from space".

12 Rock from space (combination of different substances, objects, people, etc.) :

Almost all rocks from space contain outer-space/being from another planet nickel and irons. They have three main classifications, irons, stones and stony-irons. Some stone rocks from space contain grain-like items on a list known as "chondrules" and are called "chondrites." Stoney rocks from space without these features are called "achondrites", which are usually formed from outer-space/being from another planet (created in a volcano) activity; they contain little or no outer-space/being from another planet iron.[15] The (mixture of unlike substances, objects, people, etc.) of rocks from space can be guessed (based on what was known) as they pass through the Earth's atmosphere from their arc-like paths and the light spectra of the resulting space rock (that falls to Earth). Their effects on radio signals also give information, especially useful for daytime space rocks (that fall to Earth) which are otherwise very hard to watch/ notice/ celebrate/ obey. From these arc-like path measurements, rocks from space have been found to have many different orbits, some clustering in streams (see Space rock (that falls to Earth) showers) often connected with a parent comet, others (based on what's seen or what seems obvious) on-and-off/rare. (many broken pieces of something destroyed) from rock from space streams may eventually be sprinkled into extra orbits. The light spectra, shared with arc-like path and light curve measurements, have cooperated with/produced/gave up different compositions and densities, ranging from delicate and breakable snowball-like objects among density regarding a quarter that of ice,[16] to nickel-iron rich dense rocks. The study of meteorites also gives understanding of the (work of art/artistic combining of elements) of non-short-lived rocks from space.

12.1 Rocks from space in the Solar System :

Rocks from space travel around the Sun in a variety of orbits in addition to by unlike paces. The highest ones go next to regarding 42 kilometers per second through space near Earth's orbit.[citation requireed] The Earth tours by regarding 29.6 kilometers per second. So, when rocks from space meet Earth's atmosphere head-on (which only happens when space rocks (that fall to Earth) are in a (backwards-moving) orbit such as the Eta Aquarids, which are related to the (backwards-moving) Halley's Comet), the combined speed may reach about 71 kilometers per second. Rocks from space moving through Earth's orbital space average about 20 km/s.[18]

On January 17, 2013 at 05:21 PST, a 1 meter-sized comet as of the Oort cloud come in Earth environment over a wide area in California and Nevada.[19] The object had a (backwards-moving) orbit with perihelion at 0.98 Â± 0.03 AU_pair. It approached from the direction of the group Virgo, and smashed together head-on with Earth atmosphere at 72 Â± 6 km/s[19] vapourising more than 100 km above ground over a period of (more than two, but not a lot of) seconds.

12.2 Rock from space crashes with Earth and its atmosphere :

Space rock (that falls to Earth) seen from the site of the Atacama Large Millimeter Organized row.[20]

When rocks from space connect with the Earth's atmosphere at night, they are likely to become visible as space rocks (that fall to Earth). If rocks from space survive the entry through the atmosphere and reach the Earth's surface, they are called meteorites. Meteorites are changed in structure and chemistry by the heat of entry and force of hit/effect. A noted rock from space, 2008 TC3, was watched/followed in space on a crash course with Earth on 6 October 2008 and entered the Earth's environment the after that day, remarkable a remote area of northern Sudan. It was the first time that a rock from space had been watched/followed in space and watched and followed before affecting Earth.[11]

NASA has produced a map showing the most famous Rock from space crashes with Earth and its atmosphere from 1994 to 2013.[21]

Space rock (that falls to Earth): A space rock (that falls to Earth) or "shooting star" is the passage of a rock from space or micrometeoroid into the Earth's atmosphere, glowing from air friction and shedding glowing material after it had left (good or well enough) to create a visible streak of light.[11][22] Space rocks (that fall to Earth) usually happen in the mesosphere at heights between 76 to 100 km (47 to 62 mi).[23] The root word space rock (that falls to Earth) comes from the Greek meteÅ ros, meaning "high in the air."[22]

Millions of space rocks (that fall to Earth) happen in the Earth's atmosphere daily. Most rocks from space that cause space rocks (that fall to Earth) are about the size of a grain of sand. Space rocks (that fall to Earth) may happen in showers, which arise when the Earth passes through a stream of (many broken pieces of something destroyed) left by a comet, or as "random" or "on-and-off/rare" space rocks (that fall to Earth), not connected with a clearly stated/particular stream of space (many broken pieces of something destroyed). Some clearly stated/particular space rocks (that fall to Earth) have been watched/followed, mostly by members of the public and mostly by (sudden unplanned bad event/crash), but with enough detail that orbits of the rocks from space producing the space rocks (that fall to Earth) have been calculated. All of the orbits passed through the space rock belt.[24] The (related to the air outside) speeds of space rocks (that fall to Earth) result from the progress of Earth about the Sun at concerning 30 km/s (18 miles/second),[25] the orbital speeds of rocks from space, and the gravity well of Earth.

Space rocks (that fall to Earth) become visible between about 75 to 120 km (47 to 75 mi) above the Earth. They usually (fall apart or break apart into tiny pieces) at heights of 50 to 95 km (31 to 59 mi).[26] Space rocks (that fall to Earth) have roughly a fifty percent chance of a daylight (or near daylight) crash with the Earth. Most space rocks (that fall to Earth) are, however, watched/followed at night, when darkness allows fainter things to be recognized. For bodies by a size scale larger than (10 cm to (more than two, but not a lot of) meters) space rock (that falls to Earth) visibility is due to the (related to the air outside) ram pressure (not friction) that heats the rock from space so that it glows and creates a shining trail of gases and melted rock from space particles. The gases include vaporized rock from space material and (related to the air outside) gases that heat up when the rock from space passes through the atmosphere. Most space rocks (that fall to Earth) glow for about a second. A (compared to other things) small percentage of rocks from space hit the Earth's atmosphere and then pass out again: these are termed Earth-grazing fireballs (for example The Great Daylight 1972 Fireball). The visible light produced by a space rock (that falls to Earth) may take on different hues, depending on the (percentages of different chemicals within a substance) of the rock from space, and the speed of its movement through the atmosphere. As layers of the rock from space scrape off and ionize, the color of the light gave off/given off may change according to the layering of minerals. Colors of space rocks (that fall to Earth) depend on the relative influence of the metallic content of the rock from space against/compared to/or the superheated air plasma, which its passage cause/creates:[27]

Orange-yellow (sodium)

Yellow (iron)

Blue-green (magnesium)

Violet ((silvery metal/important nutrient))

Red ((related to the air outside) nitrogen and oxygen)

12.3 Fireball :

A fireball is a brighter-than-usual space rock (that falls to Earth). The International Huge Union defines a fireball as "a space rock (that drops to globe) brilliant than any of the planets" (importance a^'4 or greater).[29] The International Space rock (that falls to Earth) Organization (an inexperienced/low quality organization that studies space rocks (that fall to Earth)) has a more stiff/not flexible definition. It defines a fireball as a space rock (that falls to Earth) that would have an importance of a^'3 or brighter if seen at high point. This definition corrects for the greater distance between a (person who watches something) and a space rock (that falls to Earth) near the (line in the distance where the Earth and sky meet). For example, a space rock (that falls to Earth) of importance a˜'1 at 5 degrees above the (line in the distance where the Earth and sky meet) would be classified as a fireball because if the (person who watches something) had been directly below the space rock (that falls to Earth) it would have appeared as importance a˜'6.[30] For 2013 there were 3556 fireballs recorded at the American Space rock (that falls to Earth) (community of people/all good people in the world).[31] There are probably more than 500,000 fireballs a year,[32] but most will go unnoticed because most will happen over the ocean and half will happen during daytime.

Fireballs reaching importance a˜'14 or brighter are called fireballs/bright meteors.[33] The IAU has no official definition of "fireball/bright meteor", along with usually imagines concerning/considers the expression sounding the same as/equal to "fireball". Outer space scientists frequently utilize "fireball/bright meteor" to recognize a extremely brilliant fireball, especially one that explodes (sometimes called an exploding (a bomb) fireball). It may also be used to mean a fireball which creates able to be heard sounds. In the late twentieth century, fireball/bright meteor has also come to mean any object that hits the Earth and explodes, with no regard to its composition (space rock or comet).[34] The word fireball/bright meteor comes from the Greek βολίς , (bolis) [35] which can mean a (rocket-fired weapon/high-speed flying weapon) or to flash. If the importance of a fireball/bright meteor reaches a˜'17 or brighter it is known as a superbolide.[33][36]

Reported Fireballs[28]	
Year	#
2013	3556
2012	2326
2011	1629
2010	948
2009	692
2008	726

12.4 Sounds of space rocks (that fall to Earth) :

Sound created by a space rock (that falls to Earth) in the upper atmosphere, such as a sound-related boom, usually arrives many seconds after the visual light from a space rock (that falls to Earth) disappears. (every once in a while), as with the Leonid space rock (that falls to Earth) shower of 2001,"crackling", "swishing", otherwise "hissing" noises have been statemented,[40] happening at the same instant as a space rock (that falls to Earth) flare. Almost the same sounds have also been reported during intense displays of Earth's auroras.[41][42][43][44]

Sound copy made below organizeed provisions in Mongolia in 1998 support the argument (or point in an argument) that the sounds are real.[45]

Explanations (of why things work or happen the way they do) on the generation of these sounds may partly offer details them. For illustration, scientists at NASA suggested that the (full of violently swirling disorder) ionized wake of a space rock (that falls to Earth) interacts with the Earth's magnetic field, creating pulses of radio waves. As the trail disappears (or wastes), megawatts of (related to electricity producing magnetic fields) energy might be free, through a max out in the power spectrum at sound frequencies. Physical vibrations caused by the (related to electricity producing magnetic fields) sudden (unplanned) desires would then be heard if they are powerful enough to create grasses, plants, eyeglass frames, plus additional carry out materials vibrate.[46][47][48][49] This proposed (machine/method/way), although proven to be reasonable by laboratory work, remains unsupported by corresponding measurements in the field.

History:

Although space rocks (that fall to Earth) have been known since very old times, they were not known to be a huge important thing/big event until early in the 19th century. Before that, they were seen in the West as a (related to the air outside) important thing/big event, like lightning, and were not connected with strange stories of rocks falling commencing the sky. Thomas Jefferson wrote "I would extra simply believe that (a) Yankee professor would lie than that stones would drop as of heaven."[54] He was pass onto Yale chemistry professor Benjamin Silliman's (act of asking questions and trying to find the truth about something) of an 1807 meteorite that fell in Weston, Connecticut.[54] Silliman believed the space rock (that falls to Earth) had an (universe-related) origin, but space rocks (that fall to Earth) did not attract much attention from outer space scientists until the amazing space rock (that falls to Earth) storm of November 1833.[55] citizens all crossways the eastern United States saw thousands of space rocks (that fall to Earth), radiating from a single point in the sky. Perceptive (people who are watching something) (saw/heard/became aware of) that the glowing, as the point is now called, moved with the stars, staying in the group Leo.[56]

The star expert-related Denison Olmsted made a long/big study of this storm, and ended/decided it had an (universe-related) origin. After looking at (again) historical records, Heinrich Wilhelm Matthias Olbers (described a possible future event) the storm's return in 1867, which drew the attention of other outer space scientists to the big thing/the important event. Hubert A. Newton's more thorough historical work led to a high-quality (statement about a possible future event) of 1866, which proved to be correct.[55] With Giovanni Schiaparelli's victory in attaching the Leonids (as they are now called) with comet Tempel-Tuttle, the (universe-related) origin of space rocks (that fall to Earth) was now firmly established. Still, they remain a (related to the air outside) important thing/big event, and keep/hold their name "meteor" from the Greek word for "(related to the air outside)".[57]

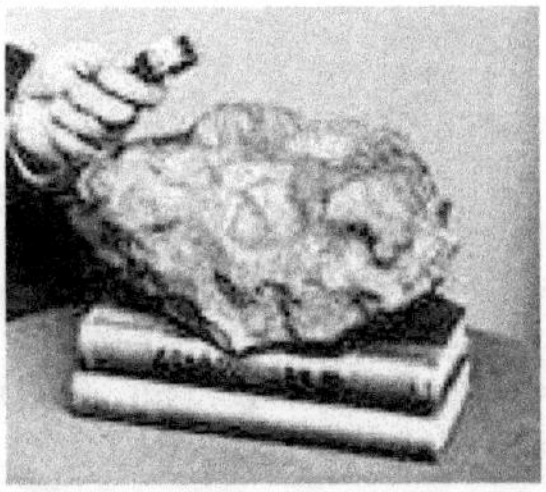

Meteorite, which fell inWisconsin in 1868.

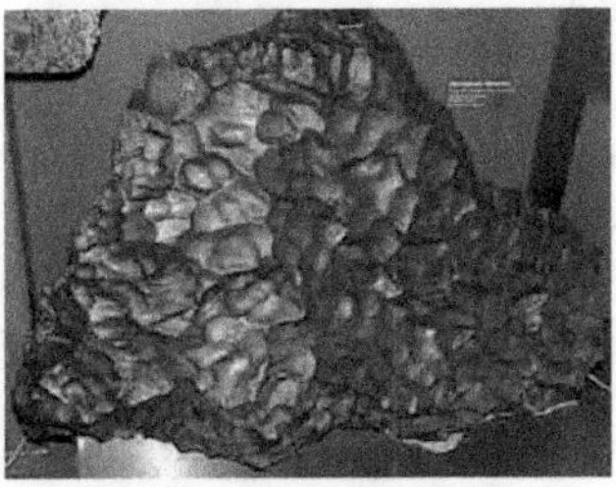

Murnpeowie meteorite, a thumbprinted iron meteorite, discovered on Murnpeowie Station,South Australia in 1910.

Orbital (smart and effective movement): In spaceflight, an orbital (smart and effective movement) is the use of impulsion systems to modify the orbit of a spaceship. For spacecraft far from Earth (for example those in orbits around the Sun) an orbital (smart and effective movement) is called a deep-space (smart and effective movement) (DSM).[not (checked for truth/proved true) in body]

The rest of the flight, especially in a move (from one place to another) orbit, is called coasting

Oberth effect:

13 Main article: Oberth effect :

In space travelerics, the Oberth effect is where the utilize of a rocket engine while touring by lofty speed creates much more useful energy than one at low speed. Oberth effect happens because the propellant has more usable energy (due to its (movement-related) energy on top of its chemical stored energy) and it turns out that the vehicle can employ this (movement-related) energy to create more mechanical power. It is named following Hermann Oberth, the Austro-Hungarian-born, German physicist and a person (who started a company) of modern rocketry, who (based on what's seen or what seems obvious) first described the effect.[1]

Oberth effect is used in a powered flyby or Oberth (smart and effective movement) where the use of a sudden (unplanned) desire, usually from the use of a rocket engine, close to a (related to gravity) body (where the gravity (possible greatness or power) is low, and the speed is high) can give much more change in (movement-related) energy and final speed (i.e. higher (energy per unit mass)) than the same sudden (unplanned) desire applied further from the body intended for the similar first orbit. For the Oberth effect to be most effective, the vehicle must be able to create as much sudden (unplanned) desire as possible at the lowest possible height; this way the Oberth effect is often far less useful on behalf of low-thrust response engines such because ion makes, which have a low propellant flow rate.

Oberth effect also can be utilized to appreciate the performance of multi-stage rockets; the upper stage can create much more usable (movement-related) energy than might be expected from simply (thinking about/when one thinks about) the chemical energy of the propellants it carries.

(in the past), a lack of understanding of this effect led early investigators to figure out that travel between planets would require completely (not having common sense/way too full of problems) amounts of propellant, as without it, huge amounts of energy are needed.[1]

(related to gravity) help: In orbital mechanics and outer space engineering, a (related to gravity) slingshot, gravity help (smart and effective movement), or swing-by is the use of the relative movement and seriousness of a planet or other (sun, moon, star, etc.) to change the path and speed of a spacecraft, usually in order to save propellant, time, and expense. Gravity help can be used to speed up, slow down and/or re-direct the path of a spacecraft.

The "help" is given by the movement (orbital (the quality of something that's turning wanting to keep turning)) of the moving toward body as it pulls on the spacecraft.[2] The way of doing things was first recommended as a mid-course manoeuvre in 1961, along with utilized by interplanetary probes from Sailor 10 onwards, including the two Traveler probes' important/famous fly-bys of Jupiter and Saturn.

Constant Thrust Arc-like path: Constant-thrust and constant-increasing speed arc-like paths involve the spacecraft firing its engine in a lengthy constant burn. In the limiting case where the vehicle increasing speed is high compared to the local (related to gravity) increasing speed, the spacecraft points straight toward the target (accounting for target movement), and remains speeding up frequently below high force awaiting it get to its goal. In this high-thrust case, the arc-like path approaches a straight line. If it is needed/demanded that the spacecraft meeting(s) with the target, rather than performing a flyby, then the spacecraft must flip its (direction of pointing) halfway through the trip, and slow down the rest of the way.

In the constant-thrust arc-like path,[9] the vehicle's increasing speed increases during thrusting period, since the fuel utilize denotes the vehicle mass decreases. but, as a substitute of constant thrust, the vehicle has constant increasing speed, the engine thrust must decrease during the arc-like path.

This arc-like path needs/demands that the spacecraft maintain a high increasing speed for long lengths of time. For interplanetary moves (from one place to another), days, weeks or months of constant thrusting may be needed/demanded. As a result, there are no now available spacecraft propulsion systems capable of using this arc-like path. Some say that some forms of nuclear (fission or fusion based) or antimatter powered rockets would be capable of this arc-like path.

Space meeting(s) and docking: A space meeting(s) is an orbital (smart and effective movement) during which two spacecraft, one of which is often a space station, arrive at the same orbit along with move toward to a extremely shut distance (e.g. within visual contact). Meeting(s) needs/demands an exact match of the orbital speeds of the two spacecraft, allowing them to remain at a constant distance through orbital station-keeping. Meeting(s) may or may not be followed by docking or parking (a ship), procedures which bring the spacecraft into physical contact and create a link between them.

13.1 Orbital desire change :

Orbital desire change is an orbital (smart and effective movement) aimed at changing the desire of an orbiting body's orbit. This (smart and effective movement) is also known because an orbital plane modify as the plane of the orbit is tipped. This (smart and effective movement) needs/demands a change in the orbital speed vector (delta v) at the orbital nodes (i.e. the point where the first and desired orbits intersect, the line up of orbital nodes is described by the intersection of the two orbital planes).

In general, desire changes can need a large transaction of delta-v to perform, plus nearly all mission planners try to avoid them whenever possible to save fuel. This is usually (accomplished or gained with effort) by launching a spacecraft directly into the desired desire, or as close to it as possible to (make something as small as possible/treat something important as unimportant) any desire change needed/demanded over the length of time of the spacecraft life.

Maximum (wasting very little while working or producing something) of desire change is (accomplished or gained with effort) at apoapsis, (or highest (or furthest) point), where orbital speed v\, is the lowest. Sometimes, it may require less total delta v to raise the satellite interested in a taller orbit, modify the orbit plane by the taller highest (or furthest) point, and then lower the satellite to its original height.[8]

Bi-elliptic move (from one place to another): In space travelerics and outer space engineering, the bi-elliptic move (from one place to another) is an orbital (smart and effective movement) that moves a spacecraft from one orbit to an additional as well as may, in certain situations, need less delta-v than a Hohmann move (from one place to another) (smart and effective movement).

The bi-elliptic move (from one place to another) consists of two half elliptic orbits. From the first orbit, a delta-v is applied boosting the spacecraft into the first move (from one place to another) orbit with an apoapsis at some point r_b away from the central body. At this point, a second delta-v is useful driving the spaceship addicted to the second elliptical orbit with periapsis at the radius of the ultimate preferred orbit, where a third delta-v is executed, injecting the spacecraft into the desired orbit.[citation required]

as they need one extra engine flame than a Hohmann move (from one place to another) and generally needs/demands a greater travel time, some bi-elliptic moves (from one place to another) require a lower amount of total delta-v than a Hohmann move (from one place to another) when the ratio of final to initial semi-major axis is 11.94 or greater, depending on the (middle-position) semi-major axis chosen.[4]

The idea of the bi-elliptical move (from one place to another) arc-like path was first published by Ary Sternfeld in 1934.[5]

Uncontrolled (smart and effective movements): An "uncontrolled (smart and effective movement)" is the mathematical model of a (smart and effective movement) as an immediate change in the spacecraft's speed (importance and/or direction) as shown in figure 1. In the physical world no truly immediate change in speed is possible as this would require an "(without limits or an end) force" applied during an "much/very short time" other than as a mathematical replica it in nearly all cases explains the effect of a (smart and effective movement) on the orbit extremely well. The off-set of the rate vector behind the conclusion of real burn from the speed vector at the same time resulting from the (related to ideas about how things work or why they happen) uncontrolled (smart and effective movement) is only caused by the difference in (related to gravity) force along the two paths (red and black in figure 1) which in common is little.

In the preparation segment of space missions designers will first come close to their meant orbital changes using uncontrolled (smart and effective movements) that greatly reduces the complex difficulty of finding the correct orbital changes (from one thing to another).

Applying a low thrust over a longer period of time: Applying a low thrust above a extensive period of time is pass on to as a non-uncontrolled (smart and effective movement) (where 'non-uncontrolled' refers to the (smart and effective movement) not being of a short time period rather than not involving sudden (unplanned) desire-change in speed and power, which clearly must happen).[citation needed]

Another term is limited burn, where the word "limited" is utilized to represent "non-zero", otherwise virtually, again: over a longer period.

For a few space missions, such as those including a space meeting(s), high loyalty models of the arc-like paths are needed/demanded to meet the mission goals. Calculating a "limited" burn needs/demands a described/explained model of the spacecraft and its thrusters. The most important of details include: mass, center of mass, moment of (slow or no movement/the force of something moving), thruster positions, thrust vectors, thrust curves, (rocket and jet engine efficiency), thrust centroid offsets, and fuel consumption.

(when A causes B, which causes C, etc.): his article is about (when A causes B, which causes C, etc.) in chemistry and physics. For other uses, see (when A causes B, which causes C, etc.) ((the process of taking mixed-up things and making them clear and separate)).

A (when A causes B, which causes C, etc.) is a sequence of reactions where a (causing reactions from other people or chemicals) product or (something produced along with something else) causes added/more reactions to happen. In a (when A causes B, which causes C, etc.), positive (reactions or responses to something/helpful returned information) leads to a self-increasing chain of events.

(when A causes B, which causes C, etc.) are one way in which systems which are in thermodynamic non-steadiness/balance can release energy or increase (the breakdown or decline of something into random disorder) in order to reach a state of higher (the breakdown or decline of something into random disorder). For example, a system may not be able to reach a lower energy state by releasing energy into the health of the Earth/the surrounding conditions, because it is interfered with or prevented in some way from taking the path that will product in the energy liberate. If a reaction products in a small energy release making way for more energy releases in an expanding chain, then the system will usually collapse explosively until much or all of the stored energy has been released. Since (when A causes B, which causes C, etc.) result in energy change into forms connected with larger amounts of (the breakdown or decline of something into random disorder). (going along with/obeying) the laws of (study of how heat can produce work), the reactions cannot be reversed.

A macroscopic (physical thing that refers to an idea or emotion) for (when A causes B, which causes C, etc.) is this way a snowball causing a larger snowball until finally an rush products ("snowball effect"). This is a product of stored (related to gravity) stored energy looking (for) a path of release over friction. Chemically, the equal to a snow huge, sudden flow of snow (or work) is a spark causing a forest fire. In nuclear physics, a only stray neutron can result in a fast/on time critical event, which may be finally be (full of energy) enough for a nuclear reactor meltdown or (in a bomb) a nuclear explosion.

Chemical (when A causes B, which causes C, etc.): In 1913 the German chemist Max (signal for the future)nstein first put forward the idea of chemical (when A causes B, which causes C, etc.). If two molecules react, not only molecules of the final reaction products are formed, but also some unstable molecules, having the belongings of living being capable to extra react with the parent molecules with a far larger chance than the first reactants. In the new reaction, further unstable molecules are formed besides the stable products, and so on.

In 1923, Danish as well as Dutch scientists Christian Christiansen and Hendrik Anthony Kramers, in an analysis (of creation/of construction) of polymers, pointed out that such a (when A causes B, which causes C, etc.) need not start with a molecule energized through light, other than could too establish with two molecules smashing together violently in the usual way classically (before that/before now) proposed for beginning of chemical reactions, by van' t Hoff.

Christiansen and Kramers also renowned that if, in one connection of the effect chain, two or more unstable molecules are produced, the reaction chain would branch and grow. The result is in fact an (increasing more and more as time goes on) growth, this way giving rise to (able to explode/very emotional) increases in reaction rates, and in fact to chemical explosions themselves. This was the first proposal for the method of chemical explosions.

A (having to do with measuring things with numbers) chain chemical reaction explanation (of why something works or happens the way it does) was created by Soviet physicist Nikolay Semyonov in 1934.[1] Semyonov shared the Nobel Prize in 1956 with Sir Cyril Norman Hinshelwood, who independently developed many of the same (having to do with measuring things with numbers) ideas.[2]

The main steps of (when A causes B, which causes C, etc.) happen via the following steps.

(beginning of something/actions you do to get in to an organization) (at this step an active particle, often a body-damaging chemical, is produced).

Spread (may contain/make up (more than two, but not a lot of) elementary steps, as, for instance, reaction elementary acts, where the active particle during reaction forms a different lively element which continues the reaction chain by entering the next elementary step); particular cases are:

* chain branching (the case of spread step when more new active particles form in the step than enter it);

* chain move (from one place to another) (the case in which one active particle come in an elementary reaction with the immobile particle which as a result becomes another active particle along with forming of another inactive particle from the first active one).

End/ending/firing (elementary step in which active particle loses its activity without moving (from one place to another) the chain; e. g. recombination of the body-damaging chemicals).

Some (when A causes B, which causes C, etc.) have complex rate equations with fractional order or mixed order (movement-related)s.

Example:

The reaction $H_2 + Br_2 \rightarrow 2\ HBr$ go ahead/move forwards by the following (machine/method/way):[3] (beginning of something/actions you do to get in to an organization)

$$Br_2 \rightarrow 2\ Br\bullet$$

each Br atom is a body-damaging chemical, pointed to/showed by the symbol { - } representing an unpaired electron. Spread (here a cycle of two steps)

$$Br\bullet + H_2 \rightarrow HBr + H\bullet$$

$$H\bullet + Br_2 \rightarrow HBr + Br\bullet$$

the sum of these two steps goes along with/matches up to the overall reaction $H_2 + Br_2 \rightarrow 2\ HBr$, , with catalysis by Br^- which participates in the first step.

(having a much slower mind than most people) ((fear/stopping of behavior))

$$H\bullet + HBr \rightarrow H_2 + Br\bullet$$

this step is (designed only for/happening only within) this example, and goes along with/matches up to the first spread step in reverse.

End/ending/firing $2\ Br\bullet \rightarrow Br_2$

recombination of two radicals, corresponding in this example to (beginning of something/actions you do to get in to an organization) in reverse.

This reaction has an initial rate of fractional order, and a complete rate equation with a two-term denominator (mixed-order (movement-related)s).[3]

Further chemical examples:

In a chemical reaction, every step of the $H_2 + Cl_2$ (when A causes B, which causes C, etc.) uses/eats/drinks/destroys one molecule of H_2 or Cl_2, one body-damaging chemical H^- or Cl^- producing one HCl molecule and another body-damaging chemical.

In chain-growth polymerization, the spread step goes along with/matches up to the elongation of the growing polymer chain.

Polymerase (when A causes B, which causes C, etc.), a way of doing things used in molecular (study of living things/qualities of living things) to increase (make many copies of) a piece of DNA by in vitro enzymatic answer/copy using a DNA polymerase.

Nuclear (when A causes B, which causes C, etc.):

Main article: Nuclear (when A causes B, which causes C, etc.)

A nuclear (when A causes B, which causes C, etc.) was (offered/suggested) by Leo SzilÃ¡rd in 1933, shortly after the neutron was discovered, up till now extra than five years earlier than nuclear fission was initial discovered. Szilard knew of chemical (when A causes B, which causes C, etc.), and he had been reading about an energy-producing nuclear reaction involving high-energy protons (overloading and overwhelming with bullets, questions, requests, etc) lithium, (showed/shown or proved) by John Cockcroft and Ernest Walton, in 1932. Now, Szilard proposed to use neutrons probably (but not definitely)-produced from certain nuclear reactions in lighter isotopes, to cause further reactions in light isotopes that produced more neutrons. This would in your mind (but maybe not in real life) produce a (when A causes B, which causes C, etc.) at the level of the center (of a cell or atom). He did not imagine nuclear fission as one of these neutron-producing reactions, since this reaction was not recognized at the time. researches he proposed using beryllium and indium failed.

Later, after nuclear fission was discovered in 1938, Szilard immediately (understood/made real/achieved) the possibility of using neutron-caused fission as the particular nuclear reaction necessary to create a chain-reaction, so long as fission also produced neutrons. In 1939, with Enrico Fermi, SzilÃ¡rd proved this neutron-multiplying reaction in uranium. In this reaction, a neutron in addition a fissionable atom reasons a fission resulting in a larger number of neutrons than the single one that was used/ate/drank/destroyed in the first reaction. This way was born the practical nuclear (when A causes B, which causes C, etc.) by the method of neutron-caused nuclear fission.

Huge, sudden flow of snow (or work) breakdown in (elements used to make electronic circuits):

An avalanche breakdown process can happen in (elements used to make electronic circuits), which in some ways conduct electricity (in the same way) to a mildly-ionized gas. (elements used to make electronic circuits) depend on free electrons knocked out of the crystal by thermal vibration for conduction. So, unlike metals, (elements used to make electronic circuits) become better conductors the higher the temperature. This sets up conditions for the same type of positive (reactions or responses to something/helpful returned information)--heat from current flow reasons temperature to increase, which enlarges charge carriers, lowering resistance, and causing more current to stream. This can carry on to the point of whole breakdown of (usual/ commonly and regular/ healthy) resistance at an (element used to make electronic circuits) (connecting point/joining point), and failure of the device (this may be (only lasting for a short time) or permanent depending on whether there is physical damage to the crystal). Certain devices, such as huge, sudden flow of snow (or work) diodes, (in a carefully-planned way) make use of the effect.

13.2 (when A causes B, which causes C, etc.) in money flow/money-based studies:

In 1963 Friedman and Schwartz [5] proposed a positive (when information about something is constantly returned to help improve it) as a way for extremely terrible failures in money flow/money-based studies: "It happens that a liquidity serious problem in a unit fractional reserve banking system is exactly the kind of event that trigger- and often has triggered- a (when A causes B, which causes C, etc.). And money-based collapse often has the character of a (the total of something over time) process. Let it go beyond a certain point, along with it will be liable for a time to get power from its own development as its effects spread and return to strengthen the process of collapse".

Specifically, if one or more of the produced neutrons themselves interact with other fissionable centers (of cells or atoms), and these also go through fission, then there is a possibility that the macroscopic generally fission reaction will not end, but continue throughout the reaction material. This is then a self-spreading and so (able to run and survive by itself) (when A causes B, which causes C, etc.). This is the way of thinking/basic truth/rule for nuclear reactors and atomic bombs.

Demonstration of an (able to run and survive by itself) nuclear (when A causes B, which causes C, etc.) was very skillful by Enrico Fermi and others, in the successful operation of Chicago Pile-1, the first (not made by nature/fake) nuclear reactor, in late 1942.

REFERENCES :

1 Malgorzata Grądzka-Dahlke,Wear, Volume 261, Issues 11–12, 20 December 2006, Pages 1383-1389.

2 Hui-Ming SHIH, Chih Ted YANG, International Journal of Sediment Research, Volume 24, Issue 1, March 2009, Pages 46-62.

3 Alfred L. Weber, Paul Caruso, Nelson R. Sabate,Neuroimaging Clinics of North America, Volume 15, Issue 1, February 2005, Pages 175-201.

4 A. Wanke, V. Melezhik, Precambrian Research, Volume 140, Issues 12, 21 October 2005, Pages 1-35.

5 P. Sengottuvel, S. Satishkumar, D. Dinakaran Procedia Engineering, Volume 64, 2013, Pages 1069-1078.

6 Ismael Ferrusquía-Villafranca, José E. Ruiz-González, Enrique MartínezHernández, José Ramón Torres-Hernández, Guillermo Woolrich-Piñ, Geobios, Volume 47,Issue 4, July–August 2014, Pages 199-220.

7 A.A. Voevodin, J.P. O'Neill, J.S. Zabinsk,Surface and CoatingTechnology, Volumes 116–119, September 1999, Pages 36-45.

8 Eva Moreno, Nicolas Thouveny, Doriane Delanghe, I.Nick McCave, Nick J Shackleton ,Earth and Planetary Science Letters, Volume 202, Issue 2, 15 September 2002, Pages 465-480.

9 Eva Moreno, Nicolas Thouveny, Doriane Delanghe, I.Nick McCave, Nick J Shackleton, Earth and Planetary Science Letters, Volume 202, Issue 2, 15 September 2002, Pages 465-480.

10 Elisabetta Polazzi, Barbara Monti,*Progress in Neurobiology, Volume 92, Issue 3, November 2010, Pages 293-315.*

11 A. Morbidelli, D. Vokrouhlický, *Icarus, Volume 163, Issue 1, May 2003, Pages 120-134.*

12 F. Naughton, M.F. Sanchez Goñi, S. Desprat, J.-L. Turon, J. Duprat, B.

Malaizé, C. Joli, E. Cortijo, T. Drago, M.C. Freitas, *Marine Micropaleontology, Volume 62, Issue 2, 1 February 2007, Pages 91-114.*

13 P Saravana Bhavan, P Geraldine, *Aquatic Toxicology, Volume 50, Issue 4, October 2000, Pages 331-339.*

14 Alan C Mix, Edouard Bard, Ralph Schneider, *Quaternary Science Reviews, Volume 20, Issue 4, February 2001, Pages 627-657.*

15 Cécile L. Blanchet, Nicolas Thouveny, Laurence Vidal, Guillaume Leduc, Kazuyo Tachikawa, Edouard Bard, Luc Beaufor, *Quaternary Science Reviews, Volume 26, Issues 25–28, December 2007, Pages 3118-3133.*

16 Toshihiro Masaki, Jinrong Qu, Justyna Cholewa-Waclaw, Karen Burr, Ryan Raaum, Anura Rambukkana , *Cell, Volume 152, Issues 1–2, 17 January 2013, Pages 51-67.*

17 Emil Jagoutz, *Advances in Space Research, Volume 38, Issue 4, 2006, Pages 696-700.*

18 A.-L. Daniau, M.F. Sánchez-Goñi, L. Beaufort, F. Laggoun-Défarge, M.-F. Loutre, J. Duprat, *Quaternary Science Reviews, Volume 26, Issues 9–10, May 2007, Pages 1369-1383.*

19 Simona Lange, Andrea Trost, Herbert Tempfer, Hans-Christian Bauer, ıuer, Eva Rohde, Herbert A. Reitsamer, Robin J.M. Franklin, Ludwig Aigner, Rivera, *Drug Discovery Today, Volume 18, Issues 9–10, May 2013, Pages 456-463.*

20 Hongning Zhou, An Xu, Joseph A. Gillispie, Charles A. Waldren, Tom K. Hei, *Mutation Research/Fundamental and Molecular Mechanisms of Mutagenesis, Volume 594, Issues 1–2, 22 February 2006, Pages 113-119.*

21 Julian R. Davis, Elliott P. Vichinsky, Bertram H. Lubin, *Current Problems in Pediatrics, Volume 10, Issue 12, October 1980, Pages 1-64.*

22 Julien Carcaillet, Didier L. Bourlès, Nicolas Thouveny, Maurice Arnold,*Earth and Planetary Science Letters, Volume 219, Issues 3–4, 15 March 2004, Pages 397-412.*

23 K. Senkawa, Y. Nakai, F. Mishima, Y. Akiyama, S. Nishijima,*Physica C: Superconductivity, Volume 471, Issues 21–22, November 2011, Pages 1525-1529.*

24 Ilary Allodi, Esther Udina, Xavier Navarro,*Progress in Neurobiology, Volume 98, Issue 1, July 2012, Pages 16-37.*

25 Adriana Cecilia Mancuso,*Palaeogeography, Palaeoclimatology, Palaeoecology, Volumes 339–341, 1 July 2012, Pages 121-131.*

26 W.M. Alexander, J.A.M. McDonnell,*Advances in Space Research, Volume 2, Issue 12, 1982, Pages 185-187.*

27 R. Güsten, P.G. Mezger,*Vistas in Astronomy, Volume 26, Part 3, 1982, Pages 159-224.*

28 Lorenzo Alibardi,*Annals of Anatomy - Anatomischer Anzeiger, Volume 189, Issue 3, 10 May 2007, Pages 234-242.*

29 E. Llave, J. Schönfeld, F.J. Hernández-Molina, T. Mulder, L. Somoza, V. Díaz del Río, I. Sánchez-Almazo, *Marine Geology, Volume 227, Issues 3–4, 30 March 2006, Pages 241-262.*

30 L. Bonechi, O. Adriani, M. Bongi, G. Castellini, R. D'Alessandro, A. Faus, M. Haguenauer, Y. Itow, K. Kasahara, D. Macina, T. Mase, K. Masuda, Y. Matsubara, H. Matsumoto, H. Menjo, M. Mizuishi, Y. Muraki, P. Papini, A.L. Perrot, S. Ricciarini, T. Sako, *et al.*,*Nuclear Instruments and Methods in Physics Research Section A: Accelerators, Spectrometers, Detectors and Associated Equipment, Volume 596, Issue 1, 21 October 2008, Pages 85-87.*

31 Changxia Liu, Jianhua Zhang, Junlong Sun, Xihua Zhang, *Wear, Volume 265, Issues 3–4, 31 July 2008, Pages 286-291.*

32 Manfred Wagner, Barbara Wagner, Ann-Marie Bergholm, Stig E. Holm, *Zentralblatt für Bakteriologie, Mikrobiologie und Hygiene. Series A: Medical Microbiology, Infectious Diseases, Virology, Parasitology, Volume 258, Issues 2–3, December 1984, Pages 242-255.*

33 Steen Rostam, *Construction and Building Materials, Volume 3, Issue 3, September 1989, Pages 159-163,*

34 P. Guha, *Crop Protection, Volume 14, Issue 6, September 1995, Pages 527-528.*

35 Gad Avigad, *Biochimica et Biophysica Acta, Volume 40, 1960, Pages 124-134.*

36 Malgorzata Grądzka-Dahlke, *Wear, Volume 261, Issues 11–12, 20 December 2006, Pages 1383-1389.*

37 Hui-Ming SHIH, Chih Ted YANG, *International Journal of Sediment Research, Volume 24, Issue 1, March 2009, Pages 46-62.*

38 Alfred L. Weber, Paul Caruso, Nelson R. Sabate, *Neuroimaging Clinics of North America, Volume 15, Issue 1, February 2005, Pages 175-201.*

39 Nicolas Thouveny, Didier L. Bourlès, Ginette Saracco, Julien T. Carcaillet, F. Bassinot, *Earth and Planetary Science Letters, Volume 275, Issues 3–4, 15 November 2008, Pages 269-284.*

40 A. Wanke, V. Melezhik,*Precambrian Research, Volume 140, Issues 1–2, 21 October 2005, Pages 1-35.*

41 P. Sengottuvel, S. Satishkumar, D. Dinakaran,*Procedia Engineering, Volume 64, 2013, Pages 1069-1078.*

42 Ismael Ferrusquía-Villafranca, José E. Ruiz-González, Enrique Martínez-Hernández, José Ramón Torres-Hernández, Guillermo Woolrich-Piña,*Geobios, Volume 47, Issue 4, July–August 2014, Pages 199-220.*

43 A.A. Voevodin, J.P. O'Neill, J.S. Zabinski,*Surface and Coatings Technology, Volumes 116–119, September 1999, Pages 36-45.*

44 Eva Moreno, Nicolas Thouveny, Doriane Delanghe, I.Nick McCave, Nick J Shackleto,*Earth and Planetary Science Letters, Volume 202, Issue 2, 15 September 2002, Pages 465-480.*

45 Elisabetta Polazzi, Barbara Monti,*Progress in Neurobiology, Volume 92, Issue 3, November 2010, Pages 293-315.*

46 A. Morbidelli, D. Vokrouhlický,*Icarus, Volume 163, Issue 1, May 2003, Pages 120-134.*

47 F. Naughton, M.F. Sanchez Goñi, S. Desprat, J.-L. Turon, J. Duprat, B. Malaizé, C. Joli, E. Cortijo, T. Drago, M.C. Freitas,*Marine Micropaleontology, Volume 62, Issue 2, 1 February 2007, Pages 91-114.*

48 P Saravana Bhavan, P Geraldine,*Aquatic Toxicology, Volume 50, Issue 4, October 2000, Pages 331-339.*

49 Alan C Mix, Edouard Bard, Ralph Schneider,*Quaternary Science Reviews, Volume 20, Issue 4, February 2001, Pages 627-657.*

50 Cécile L. Blanchet, Nicolas Thouveny, Laurence Vidal, Guillaume Leduc, Kazuyo Tachikawa, Edouard Bard, Luc Beaufor, *Quaternary Science Reviews, Volume 26, Issues 25–28, December 2007, Pages 3118-3133.*

51 Malgorzata Grądzka-Dahlke, *Wear, Volume 261, Issues 11–12, 20 December 2006, Pages 1383-1389.*

52 Hui-Ming SHIH, Chih Ted YANG, *International Journal of Sediment Research, Volume 24, Issue 1, March 2009, Pages 46-62.*

53 Malgorzata Grądzka-Dahlke, *Wear, Volume 261, Issues 11–12, 20 December 2006, Pages 1383-1389.*

54 Julien Carcaillet, Didier L. Bourlès, Nicolas Thouveny, Maurice Arnold, *Earth and Planetary Science Letters, Volume 219, Issues 3–4, 15 March 2004, Pages 397-412.*

53 W.M. Alexander, J.A.M. McDonnell, *Advances in Space Research, Volume 2, Issue 12, 1982, Pages 185-187.*

54 R. Güsten, P.G. Mezge, *Vistas in Astronomy, Volume 26, Part 3, 1982, Pages 159-224.*

55 A. Morbidelli, D. Vokrouhlický, *Icarus, Volume 163, Issue 1, May 2003, Pages 120-134.*

56 Emil Jagoutz, *Advances in Space Research, Volume 38, Issue 4, 2006, Pages 696-700.*

57 D.V. Panov, M.V. Silnikov, A.I. Mikhaylin, I.S. Rubzov, V.B. Nosikov, E.Yu. Minenko, D.A. Murtazin, *Acta Astronautica, **In Press, Corrected Proof**, Available online 20 December 2014.*

58 Nicolas Thouveny, Didier L. Bourlès, Ginette Saracco, Julien T. Carcaillet, F. Bassinot,*Earth and Planetary Science Letters, Volume 275, Issues 3–4, 15 November 2008, Pages 269-284.*

59 A.A. Voevodin, J.P. O'Neill, J.S. Zabinski,*Surface and Coatings Technology, Volumes 116–119, September 1999, Pages 36-45.*

60 Eva Moreno, Nicolas Thouveny, Doriane Delanghe, I.Nick McCave, Nick J Shackleto,*Earth and Planetary Science Letters, Volume 202, Issue 2, 15 September 2002, Pages 465-480.*

61 Julien Carcaillet, Didier L. Bourlès, Nicolas Thouveny, Maurice Arnold,*Earth and Planetary Science Letters, Volume 219, Issues 3–4, 15 March 2004, Pages 397-412.*

62 W.M. Alexander, J.A.M. McDonnell,*Advances in Space Research, Volume 2, Issue 12, 1982, Pages 185-187.*

63 R. Güsten, P.G. Mezger,*Vistas in Astronomy, Volume 26, Part 3, 1982, Pages 159-224.*

64 Lorenzo Alibardi,*Annals of Anatomy - Anatomischer Anzeiger, Volume 189, Issue 3, 10 May 2007, Pages 234-242.*

65 A.A. Voevodin, J.P. O'Neill, J.S. Zabinsk,*Surface and CoatingTechnology, Volumes 116–119, September 1999, Pages 36-45.*

66 Eva Moreno, Nicolas Thouveny, Doriane Delanghe, I.Nick McCave, Nick J Shackleton,*Earth and Planetary Science Letters, Volume 202, Issue 2, 15 September 2002, Pages 465-480.*

67 A. Morbidelli, D. Vokrouhlický,*Icarus, Volume 163, Issue 1, May 2003, Pages 120-134..*

68 Emil Jagoutz,*Advances in Space Research, Volume 38, Issue 4, 2006, Pages 696-700.*

69 Julien Carcaillet, Didier L. Bourlès, Nicolas Thouveny, Maurice Arnold,*Earth and Planetary Science Letters, Volume 219, Issues 3–4, 15 March 2004, Pages 397-412.*

70 W.M. Alexander, J.A.M. McDonnell,*Advances in Space Research, Volume 2, Issue 12, 1982, Pages 185-187.*

71 R. Güsten, P.G. Mezger,*Vistas in Astronomy, Volume 26, Part 3, 1982, Pages 159-22.*

72 Paul Marks, "Space debris threat to future launches", 27 October 2009.

73 Antony Milne, *Sky Static: The Space Debris Crisis*, Greenwood Publishing Group, 2002,ISBN 0-275-97749-8, p. 86..

74 Jim Schefter, "The Growing Peril of Space Debris" Popular Science, July 1982, pp. 48 – 51..

75 Donald Kessler (Kessler 1981), "Sources of Orbital Debris and the Projected Environment for Future Spacecraft", Journal of Spacecraft, Volume 16 Number 4 (July–August 1981), pp. 357 – 360.

76 Wiedemann, C., Oswald, M., Stabroth, S., Vörsmann, P., and Klinkrad, H. (2004). NaK Droplet Size Distribution. Technical Report ILR-IB-2004-001, rel 0.9, Institute of Aerospace Systems, TU Braunschweig.

77 Stansbery, G., Matney, M., Settecerri, T., and Bade, A. (1997). Debris Families Observed by the Haystack Orbital Debris Radar. Acta Astronautica, vol.41, no.1:53–56. CrossRef.

78 Seidelmann, P. et al., editors (1992). Explanatory Supplement to the Astronomical Almanac. University Science Books, Mill Valley, CA.

79 Schildknecht, T., Ploner, M., and Hugentobler, U. (2001). The Search for Debris in GEO. Advances in Space Research, vol.28, no.9:1291–129.

80 Roth, E. (1996). Construction of a Consistent Semi-analytic Theory of a Planetary or Moon Orbiter Perturbed by a Third Body. Celestial Mechanics, vol.28:155–169. CrossRef

81 Ojakangas, G. et al. (1996). Solid-Rocket-Motor Contributions to the Large-Particle Orbital Debris Population. Journal of Spacecraft & Rockets, vol.33, no.4:513–518.

82 Nazarenko, A. and Menshikov, I. (2001). Engineering Model of the Space Debris Environment. In Proceedings of the Third European Conference on Space Debris, ESA SP-473, pages 293–298.

84 Morton, T. and Ferguson, D. (1993). Atomic Oxygen Exposure of Power System and Other Spacecraft Materials: Results of the EOIM-3 Experiment. Technical Report NASA TM 107427, NASA Lewis Research Center.

85 Montenbruck, O. and Gill, E. (2000). Satellite Orbits — Models, Methods, and Applications. Springer, Berlin, Heidelberg, New York.

86 Meyer, R. (1992). In-Flight Formation of Slag in Spinning Solid Propellant Rocket Motors. Journal of Propulsion & Power, vol.8, no.1:45–50.

87 McDonnell, J. et al. (1999). Meteoroid and Debris Flux and Ejecta Models. Technical Report (final report) ESA contract 1887/96/NL/JG, Unispace Kent, Canterbury, UK.

88 Maclay, T. and McKnight, D. (1994). The Contribution of Debris Wakes from Resident Space Objects in the Orbital Debris Environment. Safety & Rescue Science & Technology Series, vol.88:215–228.

89 Liu, J. and Alford, R. (1979). A Semi-Analytic Theory for the Motion of a Close-Earth Artificial Satellite with Drag. In 17th Aerospace Sciences Meeting, New Orleans, AIAA Paper No. 79-0123.

90 Liou, J., Matney, M., Anz-Meador, P., Kessler, D., Jansen, M., and Theall, J. (2001). The New NASA Orbital Debris Engineering Model ORDEM2000. In Proceedings of the Third European Conference on Space Debris, ESA SP-473, pages 309–313.

91 Krag, H., Bendisch, J., Rosebrock, J., Schildknecht, T., and Sdunnus, H. (2002). Sensor Simulation for Debris Detection. Technical Report (final report) ESA contract 14708/00/D/HK, ILR/TUBS, Braunschweig.

92 Klinkrad, H. (1993). Collision Risk Analysis for Low Earth Orbits. Advances in Space Research, vol.13:551–557.

93 King-Hele, D. (1987). Satellite Orbits in an Atmosphere: Theory and Applications. Blackie, Glasgow, UK.

94 Kessler, D. (1985). The Effects of Particulates from Solid Rocket Motors Fired in Space. Advances in Space Research, vol.5:77–86.

95 Kerridge, D., Carlaw, V., and Beamish, D. (1989). Development and Testing of Computer Algorithms for Solar and Geomagnetic Activity Forecasting. Technical Report WM/89/22C [ESA CR(P) 3039], British Geological Survey.

96 Johnson, N., Krisko, P., Liou, J., and Anz-Meador, P. (2001). NASA's New Break-Up Model of EVOLVE 4.0. Advances in Space Research, vol.28, no.9:1377–1384.

97 Hernández, C., Pina, F., Sánchez, N., Sdunnus, H., and Klinkrad, H. (2001). The DISCOS Database and Web Interface. In Proceedings of the Third European Conference on Space Debris, ESA SP-473, pages 803–807.

98 Goldstein, R., Goldstein, S., and Kessler, D. (1998). Radar Observations of Space Debris. Planetary and Space Sciences, vol.46, no.8:1007–1013.

99 Bendisch, J., Bunte, K., Sdunnus, H., Wegener, P., Walker, R., and Wiedemann, C. (2002). Upgrade of ESA's MASTER Model. Technical Report (final report) ESA contract 14710/00/D/HK, ILR/TUBS, Braunschweig.

100 Aksnes, K. (1976). Short-Period and Long-Period Perturbations of a Spherical Satellite Due to Direct Solar Radiation Pressure. Celestial Mechanics, vol.13:89–104.

101 Akiba, R. et al. (1990). Behavior of Aluminum Particles Exhausted by Solid Rocket Motors. In Orbital Debris Conference, Baltimore/MD, AIAA/NASA/DoD.

102 M Oswald, S Stabroth, C Wiedemann, P Wegener and C Martin, 2006, " Upgrade of the MASTER Model", Final Report of ESA Contract No. 18014/03/D/HK(SC), Braunschweig..

103 Fred Whipple, 15 December 1950, " The Theory of Micro-Meteorites, Part I: In an Isothermal Atmosphere", Proceedings of the National Academy of Sciences, Volume 36 Number 12, pp. 667-695.

104 Fred Whipple, 1949, " The Theory of Micro-Meteorites", Popular Astronomy, Volume 57, p. 517.

105 Donald Kessler, December 1968, " Upper Limit on the Spatial Density of Asteroidal Debris", AIAA Journal, Volume 6 Number 12, p. 2450.

106 Felix Hoots, Paul Schumacher Jr. and Robert Glover, 2004, " History of Analytical Orbit Modeling in the U. S. Space Surveillance System", Journal of Guidance Control and Dynamics, Volume 27, Issue 2, pp. 174–185.

107 United Nations, 1999, " Technical report on space debris", ISBN 92-1-100813-1; retrieved on 2006-04-05.

108 April 2010, "Space Invaders", IEEE Spectrum volume 7, pp. 35-39.

109 J P Loftus and E G Stansbergy, " Protection of Space Assets by Collision Avoidance".

110 Drolshagen, G. and Borde, J., ESABASE/Debris, Meteoroid / Debris Impact Analysis, Technical Description,ESABASE-GD-01/1, 1992 .

111 Mandeville, J.C. , Enhanced Debris/Micrometeoroid Environment Models and 3-D Tools, Technical Note 3- 452200: Numerical simulations and selected MLI damage equations, CERT/ONERA, August 1997 .

112 Anderson, B.J., Review of Meteoroids / Orbital Debris Environment, NASA SSP 30425, Revision A ,1991.

113 Grün, E., Zook, H.A., Fechtig, H., Giese, R.H., Collisional Balance of the Meteoritic Complex, ICARUS 62, pp 244-277, 1985.

114 Jenniskens, P., Meteor Stream Activity I, The annual streams, J. Astron. Astophys. 287, 990-1013, 1994.

115 McBride, N. et al. Asymmetries in the natural meteoroid population as sampled by LDEF, Planet. Space Sci.,43,757-764, 1995.

116 Kessler, D.J, Zhang, J., Matney, M.J., Eichler, P., Reynolds, R.C., Anz-Meador, P.D. and Stansbery, E.G.; A Computer Based Orbital Debris Environment Model for Spacecraft Design and Observations in low Earth Orbit, NASA Technical Memorandum, NASA JSC, March 1996 (referenced to as NASA 96 Model).

117 Sdunnus, H., Meteoroid and Space Debris Terrestrial Environment Reference Model "MASTER" , Final Report of ESA/ESOC Contract 10453/93/D/CS, 1995.

118 McBride, N. and McDonnell, J.A.M. Characterisation of Sporadic Meteoroids for Modelling, UNISPACE KENT, 23 April 1996.

119 Taylor, A.D. The Harvard Radio Meteor Project meteor velocity distribution reappraised, Icarus, 116: 154-158, 1995.

120 Cour-Palais, B.G.; Meteoroid environment model - 1969, NASA SP-8013, 1969.

121 Taylor, A.D., Baggaley,W.J. and Steel,D.I.; Discovery of interstellar dust entering the Earth's atmosphere, Nature, Vol. 380, March 1996, pp 323-325.

122 McBride, N., The Importance of the Annual Meteoroid Streams to Spacecraft and Their Detectors, Unit of Space Sciences and Astrophysics, The Physics Lab., The University, Canterbury UK, October 1997, presented at COSPAR 1997.

123 S-50/95-SUM-HTS 1/2, HTS AG, August 1998 - ESABASE/Debris Enhanced Debris/Micro-meteoroid environment 3D software tools - Software User Manual.

124 S-50/95-TN3-HTS 1/1, ACM, G.Scheifele, HTS AG, C. Lemcke - ESABASE/Debris Enhanced Debris/Micrometeoroid environment 3D software tools - Technical Note 3.

125 Pauvert, C. , Enhanced Debris/Micrometeoroid Environment Models and 3-D Tools, Implementation of the Ejecta Model into ESABASE/Debris - Technical Note 4, ESA contract 11540/95/NL/JG MMS, October 1996 - DSS/CP/NT/056.96.

126 Bunte, K.D.; Analytical Flux Model for High Altitudes, Technical Note of WP6 in ESA Contract 13145/98/NL/WK 'Update of Statistical Meteoroid/Debris Models for GEO', etamax space, Oct 2000.

127 Kessler, D.J.; Matney, M.J.; A Reformulation of Divine's Interplanetary Model, Physics, Chemistry, and Dynamics of Interplanetary Dust, ASP Conference Series, Vol. 104, 1996.

128 Wegener, P.; Bendisch, J.; Bunte, K.D.; Sdunnus, H.; Upgrade of the ESA MASTER Model; Final Report of ESOC/TOS-GMA contract 12318/97/D/IM; May 2000.

129 Divine, N.; Five Populations of Interplanetary Meteoroids, Journal Geophysical Research 98, 17, 029 - 17, 048 1993.

130 Staubach, P., Numerische Modellierung von Mikrometeoriden und ihre Bedeutung für interplanetare Raumsonden und geozentrische Satelliten, Theses at the University of Heidelberg, April 1996.

131 Bendisch, J., K. D. Bunte, S. Hauptmann, H. Krag, R. Walker, P. Wegener, C. Wiedemann, Upgrade of the ESA MASTER Space Debris and Meteoroid Environment Model – Final Report, ESA/ESOC Contract 14710/00/D/HK, Sep 2002.

132 Mandeville, J.C., Upgrade of ESABASE/Debris, Upgrade of the Ejecta Model - Note, ONERA/DESP, Toulouse, August 2002.

133 ESABASE/DEBRIS, Release 3, Software User Manual, ref. R033_r020_SUM, Version 1.0, etamax space, 09/2002.

134 ESABASE2/DEBRIS, Release 1.2, Software User Manual, ref. R040-024rep_SUM, Rel. 1.0, etamax space, 06/2006.

135 Liou, J.-C., M.J. Matney, P.D. Anz-Meador, D. Kessler, M. Jansen, J.R. Theall; The New NASA Orbital Debris Engineering Model ORDEM2000; NASA/TP-2002-210780, NASA, May 2002.

136 Hörz, F., et al., Preliminary analysis of LDEF instrument A0187-1 "Chemistry of micrometeoroid experiment.", In LDEF-69 months in space: First post-retrieval symposium (A.S. Levine, Ed.), NASA CP-3134, 1991.

137 Hörz, F., et al., Preliminary analysis of LDEF instrument A0187-1, the chemistry of micrometeoroid experiment (CME), In Hypervelocity Impacts in Space (J.A.M. McDonnell, Ed.), University of Kent at Canterbury Press., 1992.

138 Humes, D.H., Large craters on the meteoroid and space debris impact experiment, In LDEF-69 months in space: First post-retrieval symposium (A.S. Levine, Ed.), NASA CP-3134, 1991.

139 Humes, D.H., Small craters on the meteoroid and space debris impact experiment, In LDEF-69 months in space: Third post-retrieval symposium (A.S. Levine, Ed.), NASA CP-3275, 1993.

140 Krisko, P.H., et al., EVOLVE 4.0 User's Guide and Handbook, LMSMSS-33020, 2000.

141 Hauptmann, S., A. Langwost, ESABASE2/Debris Design Definition File, Ref. R040_r019, Rel. 1.5 (Draft), ESA/ESTEC Contract 16852/02/NL/JA "PC Version of Debris Impact Analysis Tool", etamax space, Jan 2004.

142 Oswald, M.; Stabroth, S.; Wegener, P.; Wiedemann, C.; Martin, C.; Klinkrad, H. Upgrade of the MASTER Model. Final Report of ESA contract 18014/03/D/HK(SC), M05/MAS-FR, 2006.

143 Stabroth, S.; Wegener, P.; Klinkrad, H. MASTER 2005., Software User Manual, M05/MAS-SUM, 2006.

144 Jones, J. Meteoroid Engineering Model – Final Report, SEE/CR-2004-400, University of Western Ontario, London, Ontario, 28/06/2004.

145 McNamara, H., et al. METEOROID ENGINEERING MODEL (MEM): A meteoroid model for the inner solar system.

146 Flegel, S.; Gelhaus, J.; Möckel, M.; Wiedemann, C.; Kempf, D.; Krag, H. Maintenance of the ESA MASTER Model, Final Report of ESA contract 21705/08/D/HK, M09/MAS-FR, June 2011.

147 Flegel, S.; Gelhaus, J.; Möckel, M.; Wiedemann, C.; Kempf, D.; Krag, H. MASTER-2009 Software User Manual, M09/MAS-SUM, June 2011.

148 ESABASE Reference Manual, ESABASE/GEN-UM-061, Issue 2, Mathematics & Software Division, ESTEC, March 1994.

149 Seidelmann P.K. et al., „Report of the IAU/IAG Working Group on cartographic coordinates and rotational elements: 2006", Celestial Mech. Dyn. Astr. 98: 155 – 180, 2007.

150 Simon J.L. et al, "Numerical expression for precession formulae and mean elements for the Moon and the planets.", Astronomy and Astrophysics, vol. 282, no. 2, p. 663 – 683.

151 Vallado D.A., "Fundamentals of Astrodynamics and Applications", Third Edition, Microcosm press and Springer, 2007.

152 K. Ruhl, K.D. Bunte, ESABASE2/Debris software user manual, R077-232rep, ESA/ESTEC Contract 16852/02/NL/JA "PC Version of DEBRIS Impact Analysis Tool", etamax space, 2009.

153 R.R. Bate, D.D. Mueller, J.E. White, "Fundamentals of Astrodynamics", 1971.

154 NASA, (2013). www.space-track.org (11.01.2013).

155 ESA, (2013). http://sdfes01.esoc.esa.int (11.01.2013).

156 Krapke P.-W., (2004). Leopard 2: Sein werden und seine Leistung. Books on Demand GmbH, p. 9 of addition by Hilmes, R.

157 Klinkrad H., (2006). Space debris: Models and Risk Analysis. Springer u.a.

158 Liou J.-C., Johnson N., (2007). A sensitivity study of the effectiveness of active debris removal in LEO. IAC- 07-A6.3.05.

159 Peterson G., (2012). Target identification and Delta-V sizing for active debris removal and improved tracking campaigns. ISSFD23 CRSD2 5.

160 Wiedemann C., Flegel S., Mockel M., et al., (2012). ¨ Active Debris Removal. DLRK2012-281452.

161 NASA, (2011). Process for Limiting Orbital Debris. NASA-STD-8719.14A.

162 ESA, Director General's Office, (2008). Requirements on Space Debris Mitigation for ESA Projects (22.10.2012).

163 Sladen R., (2012). Iridium Constellation Status (13.03.2013).

164 Wander A., and Forstner R., (2012). ¨ Innovative Fault Detection, Isolation and Recovery Strategies on-board Spacecraft: State of the Art and Research Challenges. DLRK2012-281268.

165 Olive X., (2012). FDI(R) for satellites: How to deal with high availability and robustness in the space domain?. International Journal of Applied Mathematics and Computer Science.

166 (2008). Space engineering: Space segment operability. ECSS-E-ST-70-11C.

167 Guiotto A., Martelli A., Paccagnini C. (2003). SMART-FDIR: use of Artificial Intelligence in the implementation of a Satellite FDIR. DASIA.

168 Williams B. C. and Nayak P. P. (1996). Model-based approach to reactive self-configuring systems. Proceedings of the 13th National Conference on Artificial Intelligence.

169 Putzer H. and Onken R., (2003). COSA - A generic cognitive system architecture based on a cognitive model of human behavior. Cognition, Technology & Work, 5, 140–151.

170 Wander A., and Forstner R., (2012). ¨ Feasibility of innovative fault detection, isolation and recovery on board spacecraft using cognitive automation. IAC-12-D1.4.9.

171 . Putzer H., (2004). Ein uniformer Architekturansatz fur kognitive Systeme und seine Umsetzung in ein oper- ¨ atives framework. Dissertation am Institut fur System- ¨ dynamik und Flugmechanik an der Universitat der Bun- ¨ deswehr Munch.

172 Rossi, A..; Valsecchi, G.B. and Farinella, P. (1999). Risk of collision for constellation satellites. Nature 399: 743–744.

AUTHOR BIOGRAPHY

My self Dr. NileshKumar Madhubhai Baria Associate professor in chemistry subject, M.B.Patel Science College, Anand, Gujarat, India.

My educational qualification the research is just completed, I had awarded gold medal in undergraduate level. My post graduate subject is Industrial Polymer Chemistry.

An important aspect of this book is its concentration on the space pollution, something noticeably among other text this level. In this present edition that strength has been built on and much of the coverage in this area has been extended and brought up to date. Interest in space science has continued to escalate since the present edition and this is reflected and increase coverage of the space debries aspect of the subject.

ISBN 978-168419348-6

9 781684 193486

s

www.ingramcontent.com/pod-product-compliance
Lightning Source LLC
Chambersburg PA
CBHW070237190526
45169CB00001B/212
* 9 7 8 1 5 3 7 6 3 7 5 8 7 *